Zeitschrift für Angewandte Mathematik und Mechanik

ZAMM

Applied Mathematics and Mechanics

Founded by Richard von Mises in 1921

Edited at Institute of Mechanics
Academy of Sciences of the G.D.R.

THEORIA CUM PRAXI

Akademie-Verlag · Berlin

Volume 69 · 1989
Number 2

ISSN 0044-2267 ZAMM · Z. angew. Math. Mech., Berlin **69** (1989) 2, 57—112

Instructions for Authors

ZAMM publishes original papers (main articles and short communications), the lectures presented at the annual GAMM conferences, and review articles (requested by the Editorial Office of ZAMM). Papers in English (preferably) or German to be submitted for publication should be sent to the address at the bottom of this page (master and one copy), in the case of original papers together with an explicit assurance that they have not been published previously in, or are submitted to, another public journal. Authors are held fully responsible for all information included in their paper and should have one copy available for correction purposes. They get two sets of page proofs, one of them has to be returned within one week. Authors may be charged for extensive corrections in the text; corrections in the figures are not possible. They receive 150 (plenary GAMM lectures), 75 (main articles and short GAMM lectures), or 25 (short communications), respectively, free-of-charge reprints. Additional reprints have to be paid for.

Manuscripts should be type-written (one-sided, double-spaced with wide margins at both sides, each sheet numbered) beginning with the authors' names and the title of the paper. Extra types (like bold-faced letters etc.) should be marked. Please, distinguish between 1, *l*, e; *v*, ν; *w*, ω; o, O, 0 (zero); *k*, *K*, $\varkappa$; *i*, ι; *x*, $\times$ (cross), χ etc. **Formulas** must be written clearly legible (type-writer if possible, never use pencil). Formulas used in the text have to be numbered at the right margin. Use abbreviations for long terms, "exp" instead of "e" for extensive exponents, bold-faced letters for vectors etc. instead of arrows; avoid hierarchies of indices and exponents if possible. The end of a proof should be marked.

Figures should be submitted on separate sheets, drawn with Indian ink and numbered consecutively. Add figure captions, tables, and footnotes (if not avoidable) on extra sheet each.

References should be numbered in brackets throughout the text and collected according to these numbers in the appended bibliography (References). Each reference should contain author(s), title, and complete bibliographical coordinates following the subsequent examples.

Books (Reports, Preprints etc.):
IRONS, B. M.; SHRIVE N. G.: Numerical methods in engineering and applied science — Numbers are fun. Ellis Horwood Series in Mathematics and Its Applications, Ellis Horwood Ltd., Chichester 1987.

Articles in periodicals:
KASSEM, S. E.: Wave source potentials for two superposed fluids, each of finite depth. Indian J. Pure Appl. Math. **18** (1987) 2, 186—193.

Articles in Proceedings etc.:
LABISCH, F. K.: Some remarks on the morphology of non-unique solutions in nonlinear elastostatics. In: KÜPPER, T.; SEYDEL, R.; TROGER, H. (eds.): Bifurcation: analysis, algorithms, applications. Proc. Conf. at the Univ. of Dortmund, August 18—22, 1986. Internat. Ser. of Numer. Mathematics, Vol. 79, Birkhäuser Verlag, Basel-Boston-Stuttgart 1987, pp. 177—184.

To each original paper and review article, a **summary** of 5 to 10 lines in English (for manuscripts of more than 10 pages in German and Russian) and a short list of key-words and corresponding AMS (MOS) classification numbers are needed.

The precise address of each author with full name, academic title etc. is to be added. Information on address changes is appreciated.

ZAMM
Editorial Office
Rudower Chaussee 5
DDR-1199 Berlin

Hinweise für die Autoren

ZAMM veröffentlicht Originalarbeiten (Hauptaufsätze und Kleine Mitteilungen), die Vorträge der jährlichen GAMM-Tagungen und Übersichtsartikel (die von der Redaktion erbeten werden). Arbeiten in Englisch (vorzugsweise) oder Deutsch, die zur Publikation eingereicht werden, sind an die unten stehende Adresse zu senden (Original und eine Kopie). Dem Manuskript von Originalarbeiten ist eine Erklärung beizufügen, daß die Arbeit bisher in keiner anderen Zeitschrift veröffentlicht worden ist und auch kein Antrag auf Veröffentlichung in einer solchen Zeitschrift läuft. Die Autoren sind für den Inhalt ihrer Arbeiten voll verantwortlich und sollten für Korrekturzwecke eine weitere Kopie besitzen. Sie erhalten die Korrekturfahnen in zweifacher Ausfertigung; eine Korrektur ist innerhalb einer Woche zurückzusenden. Die Kosten für umfangreiche nachträgliche Korrekturen können dem Autor berechnet werden. Korrekturen in den Abbildungen sind nicht möglich. Die Autoren erhalten kostenlos 150 Sonderdrucke für die Hauptvorträge der GAMM-Tagungen, 75 für Hauptaufsätze, Übersichtsartikel und die GAMM-Kurzvorträge und 25 für Kleine Mitteilungen. Zusätzliche Sonderdrucke sind gegen Berechnung erhältlich.

Die **Manuskripte** müssen mit Maschine geschrieben sein (einseitig, zweizeiliger Zeilenabstand mit breitem Rand auf beiden Seiten, jede Seite numeriert), wobei mit den Namen der Autoren begonnen werden soll und danach der Titel folgt. Besondere Typen (z. B. halbfett) sind zu markieren. Es soll gut unterschieden werden zwischen 1, *l*, e; *v*, ν; *w*, ω, *o*, *O*, 0 (Null); *k*, *K*, $\varkappa$; *i*, ι; *x*, $\times$ (Kreuz), χ; usw. Die **Formeln** müssen deutlich lesbar sein (möglichst maschinenschriftlich, keine Verwendung von Bleistift). Im Text verwendete Formeln sollten am rechten Rand durchnumeriert werden. Für lange Terme sind Abkürzungen zu verwenden, „exp" anstelle von „e" für große Ausdrücke im Exponenten, halbfett für Vektoren anstelle von darübergesetzten Pfeilen — Mehrfach-Indizes wie auch -Exponenten möglichst vermeiden. Das Ende eines Beweises sollte gekennzeichnet werden.

Abbildungen sind auf gesonderten Blättern einzureichen und sollten mit Tusche gezeichnet und durchnumeriert sein. Die Abbildungsunterschriften, eventuelle Tabellen und (falls nicht vermeidbar) Fußnoten sind ebenfalls auf gesonderten Blättern hinzuzufügen.

Literaturzitate sind durch den gesamten Text durchzunumerieren (in eckigen Klammern) und in der entsprechenden Reihenfolge im am Schluß folgenden Literaturverzeichnis aufzulisten. Jedes Zitat sollte Autor(en), Witel und die vollständigen bibliografischen Koordinaten enthalten, entsprechend den folgenden Beispielen.

Bücher (Reports, Preprints usw.):
IRONS, B. M.; SHRIVE N. G.: Numerical methods in engineering and applied science — Numbers are fun. Ellis Horwood Series in Mathematics and Its Applications, Ellis Horwood Ltd., Chichester 1987.

Zeitschriftenartikel:
KASSEM, S. E.: Wave source potentials for two superposed fluids, each of finite depth. Indian J. Pure Appl. Math. **18** (1987) 2, 186—193.

Arbeiten in Proceedings usw.:
LABISCH, F. K.: Some remarks on the morphology of non-unique solutions in nonlinear elastostatics. In: KÜPPER, T.; SEYDEL, R.; TROGER, H. (eds.): Bifurcation: Analysis, algorithms, applications. Proc. Conf. at the Univ. of Dortmund, August 18—22, 1986. Internat. Ser. of Numer. Mathematics, Vol. 79, Birkhäuser Verlag, Basel-Boston-Stuttgart 1987, pp. 777—184.

Für jede Originalarbeit und jeden Übersichtsartikel wird eine englische **Zusammenfassung** (bei Manuskripten von mehr als 10 Seiten auch je eine in Deutsch und Russisch) von 5 bis 10 Zeilen Länge sowie eine kurze Liste von Stichworten und den entsprechenden AMS (MOS)-Klassifikationsmarken benötigt.

Die genaue Adresse jedes Autors mit vollständigem Namen, akademischen Titeln usw. muß beigefügt werden. Bei Adressenänderung wird eine umgehende Information erwartet.

Zeitschrift für Angewandte Mathematik und Mechanik

ZAMM

Applied Mathematics and Mechanics

Founded by Richard von Mises in 1921

Edited at Institute of Mechanics
Academy of Sciences of the G.D.R.

Editor-in-Chief: G. Schmidt

Akademie-Verlag · Berlin

Volume 69 · 1989
Number 2

Editorial Office

Academy of Sciences of the GDR, Institute of Mechanics
Prof. Dr. Günter Schmidt · Elke Herrmann · Dr. Dieter Ullmann · Dr. Horst Weinert

The aim and scope of ZAMM is, in agreement with the intentions of its founder Richard von Mises, one of the outstanding scientists of our century in both Mathematics and Mechanics: to publish new results and review articles, the proceedings of the annual GAMM conferences, book reviews and information on applied mathematics (mainly on numerical mathematics and various parts and applications of analysis, in particular numerical aspects of differential and integral equations), on the whole field of theoretical and applied mechanics (solid mechanics, fluid mechanics, thermodynamics) and on mathematical physics.
The journal is of interest for persons working in applied mathematics and mechanics as well as on mathematical or mechanical questions in other sciences, for instance mechanical and civil engineering, electrotechniques, and chemistry.

In Übereinstimmung mit den Zielstellungen des Begründers der ZAMM, Richard von Mises, eines der bedeutendsten Wissenschaftler unseres Jahrhunderts auf dem Gebiet der Mathematik und Mechanik, publiziert die Zeitschrift neue Ergebnisse und Übersichtsartikel, die Vorträge auf den jährlichen GAMM-Tagungen, Buchbesprechungen und Informationen über Angewandte Mathematik (hauptsächlich über numerische Mathematik und verschiedene Teile sowie Anwendungen der analysis, insbesondere numerische Aspekte bei Differential- und Integralgleichungen), über das gesamte Gebiet der theoretischen und angewandten Mechanik (Festkörpermechanik, Strömungsmehanik, Thermodynamik) und über mathematische Physik.
Die Zeitschrift wendet sich an Vertreter der angewandten Mathematik und Mechanik wie auch an Wissenschaftler und Ingenieure, die sich mit mathematischen oder theoretisch-mechanischen Fragen auf anderen Gebieten, z. B. Maschinenbau und Bauwesen, Elektrotechnik und Chemie befassen.

Terms of subscription

Orders can be sent

— **in the GDR**: to Postzeitungsvertrieb, or to Akademie-Verlag Berlin, Leipziger Str. 3—4, PF-Nr. 1233, DDR-1086 Berlin;
— **in the other socialist countries**: to a book-shop for foreign language literature or to the competent news-distributing agency;
— **in the FRG and Berlin (West)**: to a book-shop or to the wholesale distributing agency Kunst und Wissen, Erich Bieber oHG, Postfach 102844, D-7000 Stuttgart 10;
— **in the other Western European Countries**: to Kunst und Wissen, Erich Bieber GmbH, General Wille-Str. 4, CH-8002 Zürich;
— **in other countries**: to the international book- and journal-selling trade, to Buchexport, Volkseigener Außenhandelsbetrieb der DDR, Postfach 160, DDR-7010 Leipzig; or to the Akademie-Verlag Berlin, Leipziger Str. 3—4, PF-Nr. 1233, DDR-1086 Berlin.

Bezugsmöglichkeiten

Bestellungen sind zu richten:

— **in der DDR** an den Postzeitungsvertrieb oder an den Akademie-Verlag Berlin, Leipziger Str. 3—4, PF-Nr. 1233, DDR-1086 Berlin;
— **im sozialistischen Ausland** an eine Buchhandlung für fremdsprachige Literatur oder an den zuständigen Postzeitungsvertrieb;
— **in der BRD und Berlin (West)** an eine Buchhandlung oder an die Auslieferungsstelle Kunst und Wissen, Erich Bieber oHG, Postfach 102844, D-7000 Stuttgart 10;
— **in den übrigen westeuropäischen Ländern** an eine Buchhandlung oder an die Auslieferungsstelle Kunst und Wissen, Erich Bieber GmbH, General Wille-Str. 4, CH-8002 Zürich;
— **im übrigen Ausland** an den Internationalen Buch- und Zeitschriftenhandel; den Buchexport, Volkseigener Außenhandelsbetrieb der Deutschen Demokratischen Republik, Postfach 160, DDR-7010 Leipzig; oder an den Akademie-Verlag Berlin, Leipziger Str. 3—4, PF-Nr. 1233, DDR-1086 Berlin.

ZEITSCHRIFT FÜR ANGEWANDTE MATHEMATIK UND MECHANIK

Herausgeber: AdW der DDR, Institut für Mechanik, Karl-Marx-Stadt. Mit der Herausgabe beauftragt und Chefredakteur: Prof. Dr. Günter Schmidt, Rudower Chaussee 5, DDR-1199 Berlin, Fernruf: 6743669 oder 6743668 (gleichzeitig Anschrift der Redaktion).
Verlag: Akademie-Verlag Berlin, Leipziger Straße 3—4, DDR-1086 Berlin; Fernruf: 2236221 oder 2236229; Telex-Nr.: 114420; Bank: Staatsbank der DDR, Berlin, Kto.-Nr.: 6836-26-20712. Veröffentlicht unter der Lizenznummer 1282 des Presseamtes beim Vorsitzenden des Ministerrates der Deutschen Demokratischen Republik. Gesamtherstellung: VEB Druckerei „Thomas Müntzer", DDR-5820 Bad Langensalza.
Erscheinungsweise: Die Zeitschrift für Angewandte Mathematik und Mechanik erscheint monatlich. Die 12 Hefte eines Jahres einschließlich Tagungshefte bilden einen Band. Bezugspreis je Band 420,— DM zuzüglich Versandspesen. Bezugspreis je Heft 35,— DM. Der gültige Jahresbezugspreis für die DDR ist der Postzeitungsliste zu entnehmen. Bestellnummer dieses Heftes: 1009/69/2.

AN (EDV) 35937
02000

ZAMM · Z. angew. Math. Mech. 69 (1989) 2, 59–65

NOWINSKI, J. L.

Rayleigh Waves in a Half-Infinite Elastic Diatomic Space

Die Arbeit untersucht die Ausbreitung elastischer Wellen über die Oberfläche eines Halbraumes, dessen Material elastisch ist und dessen Teilchen zwei Atome verschiedener thermomechanischer Eigenschaften enthalten. Nach Aufstellung der Materialgleichungen der Bestandteile werden die Bewegungsgleichungen mit Hilfe der vier Quasi-Potentialfunktionen gelöst. Die Erfüllung der Randbedingungen ergibt die charakteristische Gleichung des Problems, welche — im Gegensatz zum konventionellen Fall und in Übereinstimmung mit seismologischen Beobachtungen — auf die Dispersion der Rayleighwellen hindeutet. Ein Beispiel mit gelockerten Bindungen zwischen den Bestandteilen wird untersucht und graphisch illustriert.

The paper examines propagation of plane waves over the surface of a half-infinite space whose material is elastic, and the particles include two atoms of different thermomechanical properties. After establishing the constitutive equations of the constituents, the equations of motion are solved by appeal to four quasi-potential functions. Satisfaction of the boundary conditions yields the characteristic equation of the problem which — in contrast to the conventional case and in agreement with seismological observations — indicates dispersion of the Rayleigh waves. An example involving relaxed bonds between the constituents is examined and illustrated by graphs.

В статье изучается распространение плоских волн через поверхность полубесконечного пространства состоящего из упругого материала. Частицы материала содержают два атома с различными термомеханическими свойствами. Установив конститутивные уравнения составных частей решаются уравнения движения при помощи четырех квази-потенциальных функций. Выполнение краевых условий дает характеристическое уравнение задачи, которое — в противоположении к обычному случаю а в согласовании с сейсмологическими наблюдениями — указывает дисперсию волн Рэлея. Рассматривается пример вовлекающий релаксационные связи между составными частями, который иллюстрируется графически.

Introduction

Whether by a coincidence of events, or by some deeper reason, solid state mechanics has been approached, in the march of time, from two diametrically opposite points of view. One line of attack committed itself to a macromodel, and treated the bodies as quasi-amorphous continua with traces of internal structure attributed to such factors as gross inhomogeneities, anisotropy, and temperature dependence of properties.

The other approach believed in the micromodel, and ascribed to the matter a discrete structure with submicroscopic corpuscles — atoms and molecules — as the building blocks. With the elapsing time, and the recent unprecedented development of science, the gap between the two divergent doctrines started to narrow.

On the one side of the aisle, in the framework of dynamics of crystal lattices established by BORN and VON KÁRMÁN [1], physicists were able to predict theoretically the elastic constants appearing in the conventional continuum theory, and to establish contacts with the latter in the range of long acoustic waves.

As to the supporters of the continuum theory, several generalizations of the classical elasto- and fluid mechanics were proposed that incorporated some kind of a microstructure into the existing macromodel. Those can be divided at least into four categories.

(1) One category, initiated by VOIGT [2] and DUHEM [3], and developed brilliantly by brothers COSSERAT [4], asserted that the interactions between particles of material continua consist not only of the conventional force-stresses but also of the couple-like *couple-stresses*. The investigations proceeded here in several directions differing somewhat on the surface, but inherently more or less related. In this connection mention should be made of the modern derivation of the Cosserat equations by TRUESDELL and TOUPIN [5], of Toupin's establishment of constitutive equations for finite deformations of polar media [6], and of MINDLIN and TIERSTEN's [7] and MINDLIN's [8] theory of linear elastic microstructure. A generalization of different avenues of approach was given by ERINGEN in the theories of micropolar bodies (admitting rigid notations of the microelements [9]) and of micromorphic bodies (acknowledging a general deformation of the microelements [10]).

(2) A theory of great generality was developed by GREEN and RIVLIN [11], [12]. These authors introduced the concepts of higher order velocities, of higher order gradients of the deformation and velocity, and of the multipolar forces and stresses as partners of the arbitrary velocity gradients. The former are defined here by the requirement that their inner products with velocities constitute the rate of working. Constitutive equations are formulated to accommodate the cases in which the multipolar forces are known, so that multipolar stresses, with some reservations, can be determined.

(3) The third line of generalization involves the so-called *nonlocalities*. While their occurrence in physical problems was occasionally noted in the past, a systematic theory of nonlocal continua based on the modern theory of constitutive equations was established in the last three decades.

The nonlocal theories come, probably, closest to the statistical-molecular formulations, inasmuch as their fundamental postulate is the existence of mutual particle interactions reaching far beyond the first neighbors. In view of this assumption, the constitutive equations of the nonlocal materials include along with the derivatives also the integrals of the functions determining the intrinsic state of the body.

The theories of the nonlocal continua were established for the most part due to the efforts of GREEN and RIVLIN [13] 1965, KROENER [14] 1966, ERINGEN [15] 1966, EDELEN [16] 1969 and KUNIN [17] 1969.

(4) In 1973 DEMIRAY [18] made known a study in which he formulated the foundations of a general continuum theory of elastic media with the so-called *diatomic* structure. In this structure the particles are composed of two interacting atoms of different thermo-mechanical properties and overlapping in their initial positions. A deformation carries them into distinct spatial places with the properties of the bulk material remaining unchanged. The theory is easily extended to *polyatomic* bodies whose particle comprise an arbitrary number of dissimilar atoms.

In the present note (see also [19]) we wish to shed some light, however limited, on the characteristic properties of a half-infinite diatomic medium transmitting elastic Rayleigh-type surface waves. We first establish the constitutive equations of the medium, and the equations of balance of the linear and angular momenta, transformed by appeal to four quasi-potential functions. The procedure yields two pairs of coupled differential equations from which the potential functions are determined. Fulfillment of the boundary conditions leads to the characteristic equation of the problem which — in contrast with the conventional case and in agreement with the observations — reveals a dispersion of Rayleigh waves. The dispersion equation is analyzed in more detail for a material in which interactions between the constituents are partially relaxed. The analysis is illustrated by graphs.

1. Equations of motion, stresses

Consider a half-infinite space referred to a cartesian rectangular coordinate frame such that the x_1x_3-plane coincides with the plane boundary, and the x_2-axis points into the interior, of the half-space. The material of the half-space is assumed to be homogeneous, isotropic, elastic and of the diatomic structure, each of the constituents being for convenience designated by the index in parentheses ((1) or (2)).

Imagine now that a plane wave propagates through the half-space in the x_1-direction, so that all kinematic and dynamic quantities involved are independent of the coordinate x_3. In the context of the theory of diatomic media derived by DEMIRAY [18], the stress components of interest in the present case become

$$\tau_{ij}^{(1)} = 2a_2e_{ij}^{(1)} + 2a_6e_{ij}^{(2)} + (a_1e_{rr}^{(1)} + a_5e_{rr}^{(2)})\,\delta_{ij} + 2a_7(\omega_{ij}^{(1)} - \omega_{ij}^{(2)}) , \tag{1.1}$$

$$\tau_{ij}^{(2)} = 2a_4e_{ij}^{(2)} + 2a_6e_{ij}^{(1)} + (a_3e_{rr}^{(2)} + a_5e_{rr}^{(1)})\,\delta_{ij} + 2a_7(\omega_{ij}^{(2)} - \omega_{ij}^{(1)}) . \tag{1.2}$$

Here $i, j = 1, 2$, the a_i's are material coefficients, δ_{ij} is the Kronecker delta,

$$e_{ij} = \tfrac{1}{2}\,(u_{i,j} + u_{j,i}) , \qquad \omega_{ij} = \tfrac{1}{2}\,(u_{i,j} - u_{j,i}) , \tag{1.3}$$

u_i, $i = 1, 2$, is the displacement component, a comma denotes the differentiation (e.g., $u_{i,j} \equiv \partial u_i/\partial x_j$), and it is summed over the repeated indices (e.g., $e_{rr} \equiv e_{11} + e_{22}$).

In the case under discussion we have for the medium (1),

$$\left.\begin{aligned} \tau_{11}^{(1)} &= (2a_2 + a_1)\,u_{1,1}^{(1)} + (2a_6 + a_5)\,u_{1,1}^{(2)} + a_1u_{2,2}^{(1)} + a_5u_{2,2}^{(2)} , \\ \tau_{22}^{(1)} &= (2a_2 + a_1)\,u_{2,2}^{(1)} + (2a_6 + a_5)\,u_{2,2}^{(2)} + a_1u_{1,1}^{(1)} + a_5u_{1,1}^{(2)} , \\ \tau_{21}^{(1)} &= (a_2 - a_7)\,u_{1,2}^{(1)} + (a_2 + a_7)\,u_{2,1}^{(1)} + (a_6 + a_7)\,u_{1,2}^{(2)} + (a_6 - a_7)\,u_{2,1}^{(2)} , \end{aligned}\right\} \tag{1.4}$$

with the corresponding stress components for the medium (2) upon using the translation displayed in the Table 1.

Table 1. Correspondence between the coefficients a_i's

Medium (1)	Medium (2)
a_1	a_5
a_2	a_6
a_5	a_3
a_6	a_4
a_7	$-a_7$

Since the displacement components are independent of the x_3-coordinate, we introduce four quasi-potential functions $\varphi_i(x_1, x_2, t)$ and $\psi_i(x_1, x_2, t)$ such that,

$$u_1^{(1)} = \varphi_{1,1} + \psi_{1,2} , \qquad u_1^{(2)} = \varphi_{2,1} + \psi_{2,2} , \qquad u_2^{(1)} = \varphi_{1,2} - \psi_{1,1} , \qquad u_2^{(2)} = \varphi_{2,2} - \psi_{2,1} , \tag{1.5}$$

and the effects of the dilatation and rotation are separated.

We now note that the equations of motion in the case under discussion are

$$\tau_{11,1}^{(i)} + \tau_{21,2}^{(i)} \pm R_1 - \varrho_1\ddot{u}_1 = 0 , \qquad \tau_{12,1}^{(i)} + \tau_{22,2}^{(i)} \pm R_2 - \varrho_2\ddot{u}_2 = 0 , \qquad i = 1, 2 , \tag{1.6}$$

where the plus sign, (minus sign) refers to the medium (1) (to the medium (2)), ϱ_1 and ϱ_2 are the mass densities of the media (1) and (2), respectively, the superposed dot denotes the time differentiation, and R_i, $i = 1, 2$, are the rates of the momentum transfer between the media. The balance of the moment of momentum implies that

$$\tau_{[ij]}^{(1)} + \tau_{[ij]}^{(2)} = 0 , \tag{1.7}$$

where

$$\tau_{[ij]}^{(i)} = \tfrac{1}{2}\,(\tau_{ij}^{(i)} - \tau_{ji}^{(i)}) \tag{1.8}$$

is the antisymmetric part of the stress tensors. It follows that in the diatomic medium, in contradistinction from the classical case, the stress tensors are asymmetric. Equation (1.7) also shows that the destabilizing effect of the asymmetry is automatically annulled by the very presence of two constituent media. On this point, the diatomic medium is at variance with the asymmetric Cosserat continuum in which the balance of asymmetric stresses is achieved by the presence of the couple-stresses themselves.

We now insert the expressions (1.5) into the equations (1.1) and (1.2), and the results into the equations (1.6). For the latter to be satisfied it is sufficient that the following system of partial differential equations holds true,

$$b_1\nabla^2\varphi_1 + b_2\nabla^2\varphi_2 + b_0(\varphi_2 - \varphi_1) = \varrho_1\ddot{\varphi}_1\,, \qquad b_2\nabla^2\varphi_1 + b_3\nabla^2\varphi_2 + b_0(\varphi_1 - \varphi_2) = \varrho_2\ddot{\varphi}_2\,, \tag{1.9}$$

$$b_4\nabla^2\psi_1 + b_5\nabla^2\psi_2 + b_0(\psi_2 - \psi_1) = \varrho_1\ddot{\psi}_1\,, \qquad b_5\nabla^2\psi_1 + b_6\nabla^2\psi_2 + b_0(\psi_1 - \psi_2) = \varrho_2\ddot{\psi}_2\,, \tag{1.10}$$

where $\nabla^2 = \partial^2/\partial x_1^2 + \partial^2/\partial x_2^2$. An inspection shows that the foregoing system of equations divides into two separate systems of pairs of coupled equations respectively in φ_1, φ_2 and ψ_1, ψ_2. The material coefficients b_i's are now defined in terms of the earlier coefficients a_i's as follows:

$$b_0 = a_0\,, \quad b_1 = a_1 + 2a_2\,, \quad b_2 = a_5 + 2a_6\,, \quad b_3 = a_3 + 2a_4\,, \quad b_4 = a_2 - a_7\,, \quad b_5 = a_6 + a_7\,, \quad b_6 = a_4 - a_7\,. \tag{1.11}$$

As to the material coefficients in general, it is convenient for the subsequent discussion, and instructive as regards the characteristic features of the diatomic media, to observe that the coefficients appearing in the constitutive equations (1.4) and in the corresponding equations for the medium (2), as well as those introduced in equations (1.8) and (2.16) to follow, may be placed into three inherently different classes displayed in Table 2.

Table 2. Correspondence between diatomic and monoatomic coefficients

Class	Diatomic coefficients	Monoatomic coefficients	
I. Constitutional coefficients	a_1	λ_1	
	a_2	μ_1	Lamé's Constants
	a_3	λ_2	
	a_4	μ_2	
II. Interaction coefficients	a_5	—	
	a_6	—	
	a_7	—	
	a_0	—	
III. Dynamic coefficients	α^2	$-k^2 + \dfrac{\varrho_1\omega^2}{\lambda_1 + 2\mu_1}$	
	β^2	$-k^2 + \dfrac{\varrho_2\omega^2}{\lambda_2 + 2\mu_2}$	
	γ^2	$-k^2 + \dfrac{\varrho_1\omega^2}{\mu_1}$	
	δ^2	$-k^2 + \dfrac{\varrho_2\omega^2}{\mu_2}$	

It is apparent from Table 2 that the diatomic coefficients of the class I play the role of the Lamé moduli of the conventional elasticity, and reduce to the latter if all of the interactions between the constituent media are severed. The interactions coefficients, that form the group II, are absent in the classical case; by suppressing them, one reduces the equations (1.4) — after a suitable translation of the constitutional coefficients — into their classical counterparts. As to the coefficients of the class III, they are shown later to correspond to the frequency terms appearing in the conventional theory.

2. Solution of the problem

We now return to the equations (1.9) and (1.10), and select their solutions in the form

$$\varphi_i(x_1, x_2, t) = F_i(x_2)\, e^{i(kx_1 - \omega t)}\,, \qquad \psi_i(x_1, x_2, t) = G_i(x_2)\, e^{i(kx_1 - \omega t)}\,. \tag{2.1}$$

where $i = 1, 2$, the letter k stands for the wave number, and ω denotes the frequency of the wave propagating in the x_1-direction. It is not difficult to decouple each of the systems (1.9) and (1.10), and after some algebra to arrive at the equations of the type,

$$F_{1,2222} + \lambda_1 F_{1,22} + \lambda_2 F_1 = 0\,, \tag{2.2}$$

where

$$\lambda_1 = \frac{b_1}{b_2\mu_1} + \nu_2 + \frac{\mu_2}{\mu_1}, \qquad \lambda_2 = \frac{b_1\nu_1}{b_2\mu_1} + \frac{\nu_2\mu_2}{\mu_1}, \tag{2.3}$$

$$\left.\begin{aligned} \mu_1 &= \frac{b_2^2 - b_1b_3}{b_2b_3(\nu_2 - \nu_4)}, \qquad \mu_2 = \frac{b_2^2\nu_2 - b_1b_3\nu_1}{b_2b_3(\nu_2 - \nu_4)}, \qquad \nu_1 = \frac{-b_1k^2 - b_0 + \varrho_1\omega^2}{b_1}, \\ \nu_2 &= \frac{-b_2k^2 + b_0}{b_2}, \qquad \nu_4 = \frac{-b_3k^2 - b_0 + \varrho_2\omega^2}{b_3}. \end{aligned}\right\} \tag{2.4}$$

In view of the commutativity property of linear operators, equation (2.2) is transformed into the pair of expressi

$$\left(\frac{d^2}{dx_2^2} - \alpha^2\right)\left(\frac{d^2F_1}{dx_2^2} - \beta^2 F_1\right) = 0\,, \tag{2.5}$$

where

$$\alpha^2 (= \tfrac{1}{2} - \lambda_1 \pm \sqrt{\lambda_1^2 - 4\lambda_2})\,, \qquad \beta^2 = \tfrac{1}{2}(-\lambda_1 \pm \sqrt{\lambda_1^2 - 4\lambda_2})\,. \tag{2.6}$$

For the sake of argument let us assume that the coefficients α^2 and β^2 are real and positive, so that α and β are real.

Furthermore, since we are looking for the type of surface waves, whose amplitude decays with an increasing value of the coordinate x_2, we simply set

$$\varphi_1 = (A_1\,e^{-\alpha x_2} + B_1\,e^{-\beta x_2})\,e^{i(kx_1 - \omega t)}\,, \tag{2.7.1}$$

where α and β have to be positive square roots of α^2 and β^2, respectively.

It is a straightforward matter to find that

$$\varphi_2 = (A_1\sigma_1\,e^{-\alpha x_2} + B_1\sigma_2\,e^{-\beta x_2})\,e^{i(kx_1 - \omega t)}\,, \tag{2.7.2}$$

where

$$\sigma_1 = \alpha^2\mu_1 + \mu_2\,, \qquad \sigma_2 = \beta^2\mu_1 + \mu_2\,. \tag{2.8}$$

Retracing the steps of the argument carried out above, it is also found that

$$\psi_1 = (A_2\,e^{-\gamma x_2} + B_2\,e^{-\delta x_2})\,e^{i(kx_1 - \omega t)}\,, \qquad \psi_2 = (A_2\sigma_3\,e^{-\gamma x_2} + B_2\sigma_4\,e^{-\delta x_2})\,e^{i(kx_1 - \omega t)}\,, \tag{2.9.1), (2.9.2}$$

where

$$\sigma_3 = \gamma^2\mu_3 + \mu_4\,, \qquad \sigma_4 = \delta^2\mu_3 + \mu_4\,, \tag{2.10}$$

$$\left.\begin{aligned} &\mu_3 = \frac{b_5^2 - b_4 b_6}{b_5 b_6(\nu_6 - \nu_7)}\,, \qquad \mu_4 = \frac{b_5^2\nu_6 - b_4 b_6 \nu_5}{b_5 b_6(\nu_6 - \nu_7)}\,, \qquad \nu_5 = \frac{-b_4k^2 - b_0 + \varrho_1\omega^2}{b_1}\,, \\ &\nu_6 = \frac{-b_5k^2 + b_0}{b_2}\,, \qquad \nu_7 = \frac{-b_4k^2 - b_0 + \varrho_2\omega^2}{b_6}\,, \end{aligned}\right\} \tag{2.11}$$

and

$$\lambda_3 = \frac{b_4}{b_5\mu_3} + \nu_6 + \frac{\mu_4}{\mu_3}\,, \qquad \lambda_4 = \frac{b_4\nu_5}{b_5\mu_3} + \nu_6\,\frac{\mu_4}{\mu_3}\,. \tag{2.12}$$

The values of γ^2 and δ^2 are found respectively from α^2 and β^2 in equations (2.6) by replacing λ_1 and λ_2 by λ_3 and λ_4 in this order. With ψ_i's and φ_i's determined, equations (1.5) yield the components of the displacements, and equations (1.4) the components of the stresses. After appeal to the Table 1 we then find

$$\tau_{22}^{(1)} = (A_1\beta_1^{(1)}\,e^{-\alpha x_2} + B_1\beta_2^{(1)}\,e^{-\beta x_2} + A_2\beta_3^{(1)}\,e^{-\gamma x_2} + B_2\beta_4^{(1)}\,e^{-\delta x_2}) \cdot e^{i(kx_1 - \omega t)}\,, \tag{2.13}$$

$$\tau_{21}^{(1)} = (A_1\beta_5^{(1)}\,e^{-\alpha x_2} + B_1\beta_6^{(1)}\,e^{-\beta x_2} + A_2\beta_7^{(1)}\,e^{-\gamma x_2} + B_2\beta_8^{(1)}\,e^{-\delta x_2}) \cdot e^{i(kx_1 - \omega t)}\,, \tag{2.14}$$

where

$$\left.\begin{aligned} \beta_1^{(1)} &= (2a_2 + a_1)\,\alpha^2 + (2a_6 + a_5)\,\alpha^2\sigma_1 - a_1k^2 - a_5k^2\sigma_1\,, \\ \beta_2^{(1)} &= (2a_2 + a_1)\,\beta^2 + (2a_6 + a_5)\,\beta^2\sigma_2 - a_1k^2 - a_5k^2\,\sigma_2\,, \\ \beta_3^{(1)} &= (2a_2 + a_1)\,\gamma ik + (2a_6 + a_5)\,\gamma ik\sigma_3 - a_1\gamma ik - a_5\gamma ik\sigma_3\,, \\ \beta_4^{(1)} &= (2a_2 + a_1)\,\delta ik + (2a_6 + a_5)\,\delta ik\sigma_4 - a_1\delta ik - a_5\delta ik\sigma_4\,, \end{aligned}\right\} \tag{2.15}$$

and

$$\left.\begin{aligned} \beta_5^{(1)} &= -(a_2 - a_7)\,ik\alpha - (a_2 + a_7)\,ik\alpha - (a_6 + a_7)\,ik\alpha\sigma_1 - (a_6 - a_7)\,ik\alpha\sigma_1\,, \\ \beta_6^{(1)} &= -(a_2 - a_7)\,ik\beta - (a_2 + a_7)\,ik\beta - (a_6 + a_7)\,ik\beta\sigma_1 - (a_6 - a_7)\,ik\beta\sigma_2\,, \\ \beta_7^{(1)} &= (a_2 - a_7)\,\gamma^2 + (a_2 + a_7)\,k^2 + (a_6 + a_7)\,\gamma^2\sigma_3 + (a_6 - a_7)\,k^2\sigma_3\,, \\ \beta_8^{(1)} &= (a_2 - a_7)\,\delta^2 + (a_2 + a_7)\,k^2 + (a_6 + a_7)\,\delta^2\sigma_4 + (a_6 - a_7)\,k^2\sigma_4\,. \end{aligned}\right\} \tag{2.16}$$

The stress components $\tau_{22}^{(2)}$ and $\tau_{21}^{(2)}$ are found from the foregoing formulas by translating the coefficients a_i's according to the Table 1.

3. Solution of the dispersion problem

At this stage we direct our attention to the boundary conditions. Assuming that the surface of the half-space is free from external tractions we have

$$\tau_{22}^{(i)}(x_1, x_2, t) = 0\,, \qquad \tau_{21}^{(i)} = (x_1, x_2, t) = 0\,, \tag{3.1}$$

for $i = 1, 2$ and $x_2 = 0$, as well as for all values of the coordinate x_1 and the time t. By inserting equations (2.13), (2.14), as well as the companion equations for the second constituent medium, into the conditions above, one arrives at a system of four homogeneous equations linear in the coefficients A_1, A_2, B_1 and B_2. Clearly the existence of a non-trivial solution of the problem requires the vanishing of the determinant of the system. This yields

$$\begin{vmatrix} \beta_1^{(1)} & \beta_2^{(1)} & \beta_3^{(1)} & \beta_4^{(1)} \\ \beta_5^{(1)} & \beta_6^{(1)} & \beta_7^{(1)} & \beta_8^{(1)} \\ \beta_1^{(2)} & \beta_2^{(2)} & \beta_3^{(2)} & \beta_4^{(2)} \\ \beta_5^{(2)} & \beta_2^{(6)} & \beta_7^{(2)} & \beta_2^{(8)} \end{vmatrix} = 0 . \tag{3.2}$$

Bearing in mind the equations (2.15) and (2.16) for the coefficients β as well as the complex structure of the equation above, there is little doubt that the characteristic equation (3.2) providing the relation between the wave number, k, and the frequency, ω (implicit in coefficients α, β, γ and δ), implies a dispersive character of the undulatory process. It follows that, in contrast to the result of the conventional elasticity, and in agreement with the seismological observations, the Rayleigh-like waves transmitted by the diatomic media are subject to dispersion. An illustrative example examined below proves that this is indeed the case even if the bounds between the constituent media are greatly relaxated.

It should be clear that, both due to its intricacy and to the presence of a great number of unknown material coefficients, a general analysis of the dispersion equation (3.2) would be of little use. In order to highlight the main points of the problem, therefore, it seems more reasonable to limit oneself either to the study: (1), of a specific class of materials, or (2), of weaker interactions between the constituent materials. In the subsequent illustrative example we select the second line of approach.

4. Illustrative example

With the decision just made in mind we limit the interactions between the constituent media to the rate of momentum transfer alone. This means that, in view of the Table 2, we set $a_5 = a_6 = a_7 = 0$, but retain $a_0 \neq 0$. It follows then from the relations (1.11) that, simultaneously, the coefficients b_2 and b_5, as well as the coefficients σ_1 and σ_3 vanish, while b_0 remains different from zero. This in turn makes the coefficients $\beta_1^{(2)}$, $\beta_3^{(2)}$, $\beta_5^{(2)}$, and $\beta_7^{(2)}$ vanish, so that the condition (3.2) automatically splits into two separate conditions:

$$[(2a_2 + a_1)\,\alpha^2 - a_1 k^2]\,(\gamma^2 + k^2) - 4k^2 \alpha\gamma a_2 = 0 , \tag{4.1}$$

associated with the medium (1), and

$$[(2a_4 + a_3)\,\beta^2 - a_3 k^2]\,(\delta^2 + k^2) - 4k^2 \beta\delta a_4 = 0 . \tag{4.2}$$

associated with the medium (2).

Here the translation of the coefficients a_i's from the equation (4.1) to the equation (4.2) agrees with their physical meaning exhibited in Table 2, group I.

We note that, due to the relaxation of bonds between the constituent media, we have

$$\left.\begin{aligned} \alpha^2 &= \frac{b_1 k^2 + b_0 - \varrho_1\omega^2}{b_1} , & \beta^2 &= \frac{b_3 k^2 + b_0 - \varrho_2\omega^2}{b_3} , \\ \gamma^2 &= \frac{b_4 k^2 + b_0 - \varrho_1\omega^2}{b_4} , & \sigma^2 &= \frac{b_6 k^2 + b_0 - \varrho_2\omega^2}{b_6} . \end{aligned}\right\} \tag{4.3}$$

If the diatomic coefficients are now converted into the Lamé elastic constants, in accordance with the Table 2 and the relations (1.11), and if the last bond between the media is severed by posing $b_0 = 0$, then either of the equations (4.1) and (4.2) reduces to the classical so-called *Rayleigh equation* (e.g., [20], eq. (2.34)). In the diatomic case under discussion, however, the equations (4.1) and (4.2) remain coupled through the medium of the coefficient b_0. Inasmuch as the formal structure of the equations (4.1) and (4.2) is identical, it suffices to analyze one of these equations. As such we select the first named equation.

We introduce the notation

$$\varkappa^2 = \frac{\varrho_1\omega^2 - b_0}{b_4 k^2} , \tag{4.4}$$

and cast equation (4.1) into the form

$$\varkappa^6 - 8\varkappa^4 + 8\left(3 - 2\frac{b_4}{b_1}\right)\varkappa^2 + 16\left(\frac{b_4}{b_1} - 1\right) = 0 , \qquad \varkappa^2 = 0 . \tag{4.5.1), (4.5.2}$$

Recalling that $\omega/k = c$ stands for the velocity of waves under discussion, and $(b_4/\varrho_1)^{1/2} = c_2$ is the velocity of the classical transverse waves, $(\mu/\varrho)^{1/2}$, the equation (4.5.2) yields

$$\frac{c^2}{c_2^2} = \frac{b_0}{b_4 k^2} , \tag{4.6}$$

as one branch of the undulatory spectrum; the branch disappears if the single existing bond between the constituent media becomes destroyed ($b_0 = 0$).

As to the equation (4.5.1), it coincides with the actual Rayleigh equation if, in accordance with the Table 2, one simply sets

$$b_4/b_1 = \mu/(2\mu + \lambda) \tag{4.7}$$

(see, e.g., [20], p. 19). If this is so, the well-known solutions of the classical equation (4.5.1) may readily be adapted to the case in question. This illustrates Fig. 1 by giving the plot of the coefficient $\varkappa$ versus the ratio b_4/b_1 varying in the interval [0, 0.5]. Clearly, the values of this coefficient associated with the real values of the coefficients α and γ, guarantee the existence of the waves in question, if they comply with the relations (4.3). This is seen if one writes the relations in the form

$$\alpha^2/k^2 = 1 - \varkappa^2 b_4/b_1 \,, \qquad \gamma^2/k^2 = 1 - \varkappa^2 \,. \tag{4.8}$$

The dependence of the square of the reduced wave velocity, c^2/c_2^2, on the square of the wave number, k^2, is depicted in the Fig. 2 for various values of b_0/b_4 and b_4/b_1.

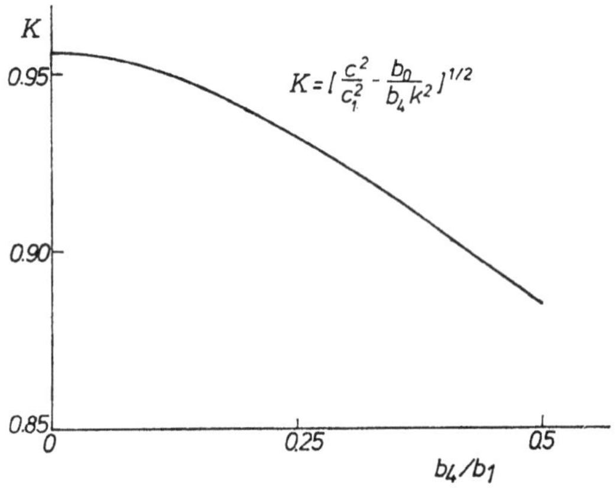

Fig. 1. Coefficient $\varkappa$ versus ratio $b_4/b_1 = a_2/(a_1 + 2a_2)$

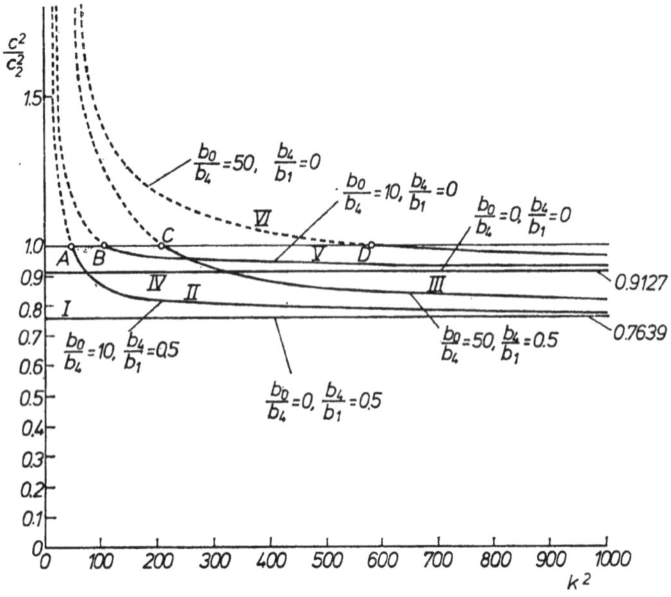

Fig. 2. Square of reduced velocity versus square of wave number

The following specific conclusions seem now to be in order:

(1) Since the analysis of the seismograms demonstrates the fact that the arrival of the Rayleigh waves takes place after the arrival of the P-waves (first) and the S-waves (second), ([21], p. 524), that is, since

$$c_R < c_2 < c_1 \,, \tag{4.9}$$

it seems reasonable to accept the value of c_2 as the upper bound for the values of the velocity of Rayleigh waves, c_R. This assumption is made apparent in Fig. 2 by the horizontal line $c^2/c_2^2 = 1$.

(2) If $b_0 = 0$, that is, if the constituent media become disrelated, then, as already noted, the dispersion of waves disappears; this fact is illustrated by the lines I and IV for $b_4/b_1 = 0.5$ and $b_4/b_1 = 0$, respectively.

(3) The curves II and III correspond to the value of $b_4/b_1 = 0.5$, and the values of $b_0/b_1 = 10$ and 50, respectively. Likewise, the curves V and VI are associated with the value of $b_4/b_1 = 0$ and the values of $b_4/b_1 = 10$ and 50, respectively. In all of these cases the wave motion is dispersive, and tends asymptotically to its classical counterpart; this takes place for the value of k^2 increasing, that is, for the wave length Λ decreasing (recall that $\Lambda = 2\pi/k$).

(4) In view of the cut-off assumption $c \leqq c_2$, the portions of the curves II, III and V, VI located above the line $c/c_2 = 1$ correspond to the imaginary values of the coefficient γ, and are not associated with any physically meaningful non-attenuating motion. Still a wave propagation may exist in a limited range of k if the first of the relations (4.8) is satisfied, that is, if the ratio b_4/b_1 is sufficiently small.

(5) The points A, B, C, and D on the line $c^2/c_2^2 = 1$ mark the value of k^2 corresponding to the cut-off wave velocities. For the ratio b_4/b_1 fixed, the cut-off velocities occur at some higher values of k^2 if the value of the ratio b_0/b_4 increases. For the ratio b_0/b_4 fixed, the same phenomenon is observed if the ratio b_4/b_1 decreases.

(6) All of the preceding conclusions are true for any diatomic material provided that the bonds are suitably relaxed.

Acknowledgement: The paper was prepared with the partial support of the University of Delaware. The author expresses his thanks to his wife, Maria Kirchmayer-Nowinski, for her cooperation in preparing this paper.

References

1 BORN, M.; KUN HUANG: Dynamical theory of crystal lattices. Clarendon Press, Oxford 1968.
2 VOIGT, W.: On media without inner forces and implied by them mechanical sense of Maxwell-Hertz equations (in German), Abh. Ges. Wiss. **34** (1894), 72–79.
3 DUHEM, P.: Thermodynamic potential and hydrostatic pressure (in French). Ann. Ecole Norm. **10** (1893), 187–230.
4 COSSERAT, E.; F.: Theory of deformable bodies (in French), Hermann, Paris 1909.
5 TRUESDELL, C.; TOUPIN, R. A.: The classical field theories. Encyclopedia of Physics, Vol. **III/I**, Sect. 200, 203, 205, Berlin, Springer-Verlag 1960.
6 TOUPIN, R. A.: Elastic materials with couple-stresses. Arch. Rat. Mech. Analysis **2** (1962), 385–414.
7 MINDLIN, R. D.; TIERSTEN, H. F.: Effects of couple-stresses in linear elasticity. Arch. Rat. Mech. Analysis **2** (1962), 415–448.
8 MINDLIN, R. D.: Micro-structure in linear elasticity. Arch. Rat. Mech. Analysis **16** (1964), 51–78.
9 ERINGEN, A. C.; Theory of micropolar continua. In: Proceedings of the 9th Midwestern Mechanics Congress, Vol. **3**, New York, Wiley 1967, pp. 23–40.
10 ERINGEN, A. C.: Mechanics of micromorphic continua. In: KROENER, E. (ed.): Mechanics of generalized continua. New York, Springer-Verlag 1968, pp. 18–35.
11 GREEN, A. E.; RIVLIN, R. S.: Simple force and stress multipoles. Arch. Rat. Mech. Analysis **16** (1964), 325–353.
12 GREEN, A. E.; RIVLIN, R. S.: Multipolar continuum mechanics. Arch. Rat. Mech. Analysis **17** (1964), 113–147.
13 GREEN, A. E.; RIVLIN, R. S.: Multipolar continuum mechanics: Functional theory I. Proc. Roy. Soc. London, Ser A **248** (1965), 303–324.
14 KROENER, E.: Continuum mechanics and range of atomic cohesion forces. In: Proc. Internat. Conf. on Fracture, Vol. **1**, Japanese Society for Strength for Fracture of Materials 1969, p. 27.
15 ERINGEN, A. C.: Nonlocal polar field theories. In: ERINGEN, A. C. (ed.): Continuum physics. Vol. **4**, Academic Press, New York 1976, pp. 205–267.
16 EDELEN, D. G. B.: Nonlocal field theories. In: ERINGEN, A. C. (ed.): Continuum physics. Vol. **4**, Academic Press, New York 1976, pp. 75–204.
17 KUNIN, I. A.: Elastic media with microstructure. Vols. **1** and **2**. Springer-Verlag, Berlin 1982 and 1983.
18 DEMIRAY, H.: A continuum theory of elastic solids with diatomic structure. Internat. J. Engineering Sci. **11** (1973), 1237–1246,
19 NOWINSKI, J. L.: Wave propagation in a circular rod made of an elastic diatomic material. Internat. J. Enginnering Sci. **25** (1987). 1165–1173.
20 KOLSKY, H.: Stress waves in solids, Dover, New York 1963.
21 ERINGEN, A. C.; SUHUBI, E. S.: Elastodynamics. Vol. 2: Linear theory, Academic Press, New York 1975.

Received January 27, 1988

Address: Dr. JERZY L. NOWINSKI, H. Fletcher Brown Professor Emeritus, Department of Mechanical Engineering, University of Delaware, Newark, Del. 19716, USA

BOOK REVIEWS

blick über moderne Herangehensweisen an aktuelle Aufgabenstellungen.

Karl-Marx-Stadt H. GÜNTHER

Bevilacqua, L.; Feijóo, R.; Valid, R. (eds.), Inelastic Behaviour of Plates and Shells. IUTAM Symposium, Rio de Janeiro, Brazil. Berlin etc., Springer-Verlag 1986. XXII, 652 S., DM 198,—. ISBN 3-540-16366-2

Der Tagungsband enthält 31 Beiträge, die in der Mehrzahl Probleme der Modellierung und Berechnung von Platten- und Schalenkonstruktionen inelastischer Materialien behandeln.

Aktuellen Aufgaben der modernen Bautechnik entsprechend, werden schwerpunktmäßig folgende Fragestellungen erfaßt: Thermische Belastung durch hohe Temperaturen, Schadensfälle (Bruch, Riß, Ermüdung u. a.), Stabilitätsfragen einschließlich des Beul- und Nachbeulverhaltens. Das breite Spektrum der behandelten Aufgabenstellungen und Methoden kann an dieser Stelle nur durch eine Auswahl der durchweg interessanten Artikel angedeutet werden.

DESTUYNDER und NGUYEN verwenden die Methode der asymptotischen Entwicklung zur Ableitung von zweidimensionalen elastisch-plastischen Platten- und Schalenmodellen aus der dreidimensionalen Aufgabenformulierung.

SZYSZKOWSKI und GLOCKNER beschreiben das zeitabhängige Verhalten von Platten aus Eis durch ein nichtlineares viskoelastisches Konstitutivgesetz. Eine angegebene inkrementelle Formulierung dieses Modells eignet sich besonders gut für eine numerische Lösung.

Einen Überblick über den modernen Stand der Theorien zum Bruch und zur Materialermüdung für Platten und Schalen aus Kompositwerkstoffen enthält der Beitrag von BOLOTIN. Schwerpunktmäßig werden Brucherscheinungen und das Abreißen von Materiallagen (Delamination) geschichteter Werkstoffe unter verschiedenen Beanspruchungen untersucht.

COCKS und LECKIE betrachten zyklisch belastete Schalen unter Berücksichtigung mikromechanischer Gesetze. Besonderes Interesse gilt dabei dem Zusammenhang von Hochtemperatur-Deformation und verschiedenen Materialschädigungen.

Einen Überblick über verschiedene Variationsmethoden zur Untersuchung des Beul- und Nachbeulverhaltens gibt der Artikel von VALID. Dabei werden sowohl dünne elastische als auch dünne plastische Schalen behandelt.

Neben anderen Autoren beschreibt BANICHUK Probleme des optimalen Entwurfs für Strukturen aus elastisch-plastischem Material. Die damit verbundene Sensitivitätsanalyse erfolgt durch die in letzter Zeit stark an Bedeutung gewinnende „Augmented Lagrange" Methode.

Im Beitrag von BALTOV wird die von diesem Autor entwickelte und auf der mathematischen Systemtheorie basierende Finite System Methode vorgestellt. Deren Eignung zur Berechnung großer Durchbiegungen viskoplastischer Platten wird an numerischen Beispielen demonstriert.

MANG und Coautoren stellen ein zweiachsiges Festigkeitskriterium für Beton vor und diskutieren dessen Verwendbarkeit bei der Traglastermittlung von Stahlbetonschalen und deren Analyse mittels FEM.

WUNDERLICH und SPRINGER leiten ein elastisch-plastisches Schalenmodell ab, das auf der Verwendung gemischtvarianter Komponenten der Spannungs- und Verzerrungstensoren beruht.

Einen Überblick über klassische und nichtklassische Vorgehensweisen bei der Shakedown-Analyse dünner Rohre wird von MAHRENHOLTZ und Coautoren gegeben. Neben plastischem Materialverhalten werden geometrische Nichtlinearitäten und Temperatureinflüsse untersucht.

Der vorliegende Tagungsband ist für den Theoretiker und für den Praktiker von Interesse. Er ermöglicht einen guten Über-

Mura, T., Micromechanics of Defects in Solids. Second, revised edition. Dordrecht etc., Martinus Nijhoff Publishers 1987. XIII, 587 pp., US $ 162.—/Dfl. 345.—. ISBN 90-247-3343-X (Mechanics of Elastic and Inelastic Solids 3)

Es gibt bisher nur sehr wenige Monografien zur Mikromechanik von Defekten in Festkörpern. Der von T. MURA verfaßte Band 3 der Reihe „Mechanik elastischer und inelastischer Festkörper", dessen zweite bearbeitete Auflage nun vorliegt, gehört inzwischen zu den international bedeutendsten Schriften auf diesem Fachgebiet.

Die sieben Kapitel sind grundlegenden Fragestellungen der „Kontinuumsmechanik" der Defekte und deren Wechselwirkungen gewidmet. Ausgehend von einem schönen Überblick über die Theorie inkompatibler Verzerrungszustände befaßt sich der Autor mit den wichtigsten mathematisch-physikalischen Modellen der Kontinuumstheorie der Defekte.

Der Anwendung der Theorie auf isotrope und anisotrope Einschlüsse sowie ellipsoidförmige Inhomogenitäten sind drei umfangreiche Kapitel des Werkes gewidmet. Der internationale Stand auf diesem Gebiet wird umfassend ausgewertet. Der Verfasser hat selbst maßgeblich zur Entwicklung beigetragen. Der Abschnitt über Risse ist angesichts der stürmischen Entwicklung der Bruchmechanik gerade in den letzten Jahren etwas zu kurz und beschränkt sich nur auf elastische Modelle, da allgemeinere Probleme schwer in das Grundkonzept der Monografie einfügbar sind.

Das Kapitel zu den Versetzungen muß unter denselben einschränkenden Voraussetzungen referiert werden. Es enthält sehr interessante und für den Anwender wichtige Gesichtspunkte der Kontinuumstheorie der Versetzungen. Es kann aber nicht als repräsentativ für die aktuellen Forschungsarbeiten zur „Versetzungstheorie" generell angesehen werden, die heute doch stärker an den Erkenntnissen der modernen Festkörperphysik orientiert sind.

Die 20 Seiten zur Plastizität als Teilkapitel dieser Problematik mögen die obige Aussage etwas belegen. Um Mißverständnisse auszuschließen, sei jedoch bemerkt, daß die behandelten Modelle von großem Interesse sind, originelle Lösungsmethoden verdeutlichen, aber eben nur spezielle Fragen der in zahlreichen eigenständigen Monografien abgehandelten Thematik „Versetzungen und Plastizität" zur Diskussion bringen können. Auf den verbleibenden etwa 100 Seiten gelingt es dem Autor gut, eine ganze Reihe von interessanten und für den Anwender aus Naturwissenschaft und Technik wichtigen Applikationsbeispielen der Theorie zu behandeln. Der Terminus „Werkstoffmechanik" wird für diese Betrachtungsweise auch international zunehmend gebräuchlich. Die neuen Werkstoffe des „high tech" erfordern derartige anspruchsvollere Modelle der Mikromechanik. Seien es nun die Composite, Keramiken oder beschichtete Sonderwerkstoffe, um nur einige wichtige Anwendungen zu nennen.

Das Werk von MURA ist ein sehr schönes Buch, das sowohl für die Forschung als auch für die Lehre an Universitäten und Hochschulen gut geeignet ist. Das ausführliche Quellenverzeichnis und die für den Studierenden hilfreichen Anhänge des Bandes sind auch unter diesem Gesichtspunkt sehr zu begrüßen. Es kann durchaus davon ausgegangen werden, daß T. MURAS Buch zur Mikromechanik der Defekte in den Bibliotheken der Lehr- und Forschungseinrichtungen der Mechanik aber auch darüber hinaus eine recht begehrte Monografie sein wird.

Karl-Marx-Stadt B. MICHEL

ZAMM Z. angew. Math. Mech. 69 (1989) 2, 67 – 71

PHAN-THIEN, N.; CHEW, Y. T.

On a Finite Deformation of a Cone

Die vorliegende Arbeit befaßt sich mit einer symmetrischen, endlichen Verformung eines elastischen Kegels. Es wird gezeigt, daß sich das Problem auf ein Zwei-Punkt-Randwertproblem reduziert. Dies wird mittels einer Finiten-Differenzen-Methode gelöst, die sich einer Zweig-Verfolgungstechnik bedient. Es wird festgestellt, daß die Lösung in einem Limit-Punkt auf sich selbst zurückklappt und dahinter ihre Stabilität verliert.

This note is concerned with a symmetric finite deformation of an elastic cone. We show that the problem reduces to a two-point boundary-value problem. This is solved by a Finite Difference method using a branch following technique. It was found that the solution folds back on itself at a limit point and losses its stability across there.

Настоящая работа посвящена с симметричной, конечной деформацией упругого конуса. Доказывается, что задача понижается на двухточечную краевую задачу. Эта задача решается при помощи конечно-разностного метода, где применяется метод предследования ветви. Наидется, что решение отвернует на себе в предельной точке и что оно теряет устойчивость посади этой точки.

1. Introduction

In a recent work [1], we have considered the finite deformation of an elastic wedge, and it was shown that, at a given wedge angle, the solution folds back on itself at a limit point and losses its stability there. In this paper, we show that the axi-symmetric deformation of a cone also has a similarity solution, and the problem is reduced to solving a nonlinear boundary-value problem. Although the kinematics are admissible for any elastic solid, we only consider the MOONEY material [2]. The resulting boundary-value problem is solved by a Finite Difference method using a branch following technique described in KUBÍČEK and MAREK [3]. A limit point also occurs in this problem; at a semi-cone angle of 45 deg in the deformed configuration, this limit point corresponds to squeezing the cone from a semi-cone angle of 55.3 deg to 45 deg, when the material is neo-Hookean.

2. Analysis

We consider the axi-symmetric finite deformation of a cone of semi-cone angle a, as illustrated in Fig. 1. In its undeformed configuration the cone occupies a region in $\mathfrak{R}^3$ described in spherical coordinates by

$$\mathcal{D}_0 = \{(R, \Phi, \Theta): 0 \leqq R, 0 \leqq \Phi < 2\pi, 0 \leqq \Theta \leqq A\},$$

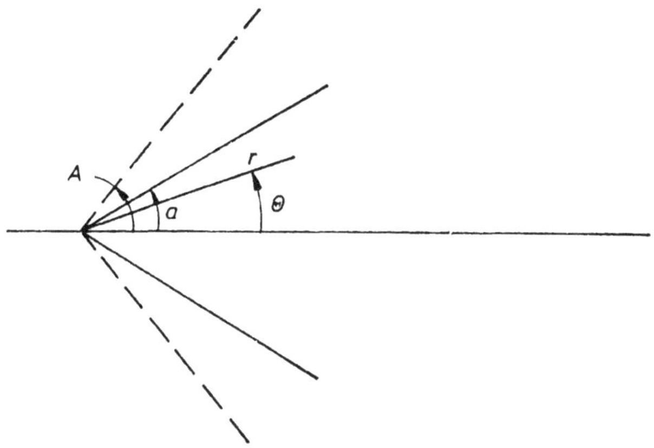

Fig. 1. Deformed configuration of the cone

where R is the radial coordinate measured from the apex of the cone, Θ is the angle measured from the axis of symmetry of the cone and Φ is the angle measured from any axis orthogonal to the axis of symmetry of the cone. After deformation, $\mathcal{D}_0$ is mapped into a region $\mathcal{D}$ described in spherical coordinates by

$$\mathcal{D} = \{(r, \varphi, \theta): 0 \leqq r, :0 \leqq \varphi < 2\pi, 0 \leqq \theta \leqq a\},$$

where a is the semi-cone angle in its deformed state. Note that we use capital letters to denote quantities in the undeformed configuration and small letters in the deformed configuration. We define the *normalized polar angle* by

$$x = \theta/a, \tag{1}$$

which takes values in the interval [0, 1]. Furthermore, we define Δ by

$$A = a(1 + \Delta). \tag{2}$$

Thus Δ can be thought of as a *measure of the deformation*. We consider the following deformation

$$R = rf(x) , \qquad \Theta = ag(x) , \qquad \Phi = \varphi , \tag{3}$$

which maps $\mathcal{D}$ into $\mathcal{D}_0$. If (3) is an admissible deformation field, it must satisfy the equilibrium equations and the incompressibility constraint (we assume the material is incompressible). Now, the *inverse displacement gradient tensor* is,

$$[F^{-1}] = \begin{bmatrix} f & f/'a & 0 \\ 0 & fg' & 0 \\ 0 & 0 & \dfrac{\sin ag}{\sin ax} f \end{bmatrix} , \tag{4}$$

where the prime denotes a derivative with respect to x. The incompressibility constraint is satisfied identically if $\det F = 1$, that is

$$f^3 = \frac{\sin ax}{\sin (ag)\, g'} . \tag{5}$$

Furthermore, from (4) we can calculate the components of the left Cauchy-Green tensor and its inverse. They are given respectively by

$$[B] = [FF^T] = \begin{bmatrix} \dfrac{1}{f^2} + \dfrac{f'^2}{a^2 f^4 g'^2} & -\dfrac{f'}{af^3 g'^2} & 0 \\ -\dfrac{f'}{af^3 g'^2} & \dfrac{f}{f^4 g'^2} & 0 \\ 0 & 0 & \dfrac{\sin^2 ax}{\sin ag} \dfrac{1}{f^2} \end{bmatrix} , \tag{6}$$

and

$$[B^{-1}] = [F^{-T}F^{-1}] = \begin{bmatrix} f^2 & \dfrac{1}{a} ff' & 0 \\ \dfrac{1}{a} ff' & \dfrac{1}{a^2} f'^2 + f^2 g'^2 & 0 \\ 0 & 0 & \dfrac{\sin^2 ag}{\sin^2 ax} f^2 \end{bmatrix} . \tag{7}$$

Since the strain tensors are functions of θ, the stress tensor for an elastic material must necessarily be a function of θ alone. We write the stress tensor as

$$\sigma = -P\mathbf{1} + \mathbf{S} , \tag{8}$$

where P is the hydrostatic pressure and $\mathbf{S}$ is the "extra" stress; both are functions of θ. Without specifying a constitutive equation for $\mathbf{S}$, we note that the equations of equilibrium read, in this axi-symmetric deformation field:

$$\frac{\partial P}{\partial r} = \frac{1}{r} \left\{ \frac{1}{\sin \theta} \frac{\mathrm{d}}{\mathrm{d}\theta} (S_{r\theta} \sin \theta) + 2S_{rr} - S_{\theta\theta} - S_{\varphi\varphi} \right\} , \qquad \frac{\partial P}{\partial \theta} = 3S_{r\theta} + \frac{1}{\sin \theta} \frac{\mathrm{d}}{\mathrm{d}\theta} (S_{\theta\theta} \sin \theta) - S_{\theta\theta} \cot \theta . \tag{9), (10}$$

An integral of (9—10) is

$$P = c \ln r + d(\theta) , \tag{11}$$

where c is a constant and d is a function of θ such that

$$\frac{1}{\sin \theta} \frac{\mathrm{d}}{\mathrm{d}\theta} (S_{r\theta} \sin \theta) + 2S_{rr} - S_{\theta\theta} - S_{\varphi\varphi} = c , \qquad \frac{\mathrm{d}}{\mathrm{d}\theta} d = 3S_{r\theta} + \frac{1}{\sin \theta} \frac{\mathrm{d}}{\mathrm{d}\theta} (S_{\theta\theta} \sin \theta) - S_{\theta\theta} \cot \theta . \tag{12), (13}$$

Thus, since (3) satisfies the incompressibility constraint, and that a scalar pressure can be found to balance the equations of equilibrium, the deformation (3) is admissible. Note also that no constitutive equation has been specified, and therefore (3) is "universal". The problem reduces to solving the boundary-value problem (12), subjected to

$$f'(0) = 0 , \qquad g(0) = 0 , \qquad f(1) = 1 , \qquad g(1) = 1 + \Delta . \tag{14}$$

The boundary conditions at $x = 0$ are recognized as the symmetry boundary conditions. The boundary conditions at $x = 1$ are the customary no slip boundary conditions. The physical situation corresponds to squeezing the cone from a semi-cone angle $A = a(1 + \Delta)$ to a semi-cone angle a. Other types of boundary are possible, but not considered here.

Suppose the edge effect is negligible and that the height of the cone is h. Then the applied torque in the deformed configuration is

$$\Gamma = \int_0^h 2\pi(P - S_{\theta\theta}) \frac{z^2 \tan a}{\cos^2 a} \, \mathrm{d}z .$$

The leading term of this is

$$\Gamma = 2\pi \frac{\tan a}{\cos^2 a} \operatorname{ch}^3(\ln h + O(1)) . \tag{15}$$

Thus c can be regarded as a normalized load on the deformed cone.

Although the kinematics (3) are "universal", we consider only the response of a Mooney material where

$$\boldsymbol{S} = \alpha \boldsymbol{B} - \beta \boldsymbol{B}^{-1} , \tag{16}$$

in which α and β are constant elastic moduli. The model describes rubber-like materials qualitatively well. We define

$$\varepsilon = \beta/\alpha , \tag{17}$$

where ε is of the order 0.3.

We can now calculate the stress components in (12) and eliminate the constant c to obtain

$$\begin{aligned} &13g'^2fff'' - 20g'^2f'^3 - 16g'g''ff'^2 - g'^2f^2f''' + 4g'g''f^2f'' + 2g'g'''f^2f' - 6g''^2f^2f' - 6ff'^2g'g'' - \\ &- 4a^2g'^4f^2f' + 2a^2g'^2f^2f' + 2a^2g'g''f^3 - 4a^2f^8f'g'^6 - 2a^2f^9g''g'^5 - a\cot(ax)\, f^2f''g'^2 + \\ &+ a^2(1 + \cot ax)\, f^2f'g'^2 + 3a\cot(ax)\, ff'^2g'^2 + 2a\cot(ax)\, f^2f'g'g'' + \\ &+ \varepsilon g'^4f^5 \left\{ -4a^2ff' - f'f'' - ff''' + 2a^2g'^2ff' + 2a^2g'g''f^2 - \frac{4a^2f'}{f^5g'^2} - \frac{2a^2g''}{f^4g'^3} - a\cot(ax)f'^2 - \right. \\ &\qquad \left. - a\cot(ax)\, ff'' + a^2(1 + \cot ax)\, ff' \right\} = 0 . \end{aligned} \tag{18}$$

This equation together with the incompressibility constraint (5) and the boundary conditions (14) form a 4th-order boundary-value problem. From (12), we find that

$$C \equiv a^2c = \left[-\frac{f''}{f^3g'^2} + a^2 \left\{ \frac{2}{f^2} - \frac{1}{f^2g'^2} - f^4g'^2 \right\} + \varepsilon \left\{ -ff'' + a^2 \left(-2f^2 + f^2g'^2 + \frac{1}{f^4g'^2} \right) \right\} \right]_{x=0} . \tag{19}$$

3. Numerical method

The resulting non-linear boundary-value problem is solved by a Finite Difference method. We divide the interval $0 \leqq x \leqq 1$ into $M + 1$ sub-intervals, each of equal length $\delta = 1/(M + 1)$. At the nodal point $x = j\delta$, we approximate derivatives by a central finite difference scheme (with a discretization error of $O(\delta^2)$):

$$f_j' = \frac{f_{j+1} - f_{j-1}}{2\delta} , \qquad f_j'' = \frac{f_{j+1} - 2f_j + f_{j-1}}{\delta^2} , \qquad f_j''' = \frac{f_{j+2} - 2f_{j+1} + 2f_{j-1} - f_{j-2}}{2\delta^3} ,$$

where f_j is the value of $f(x_j)$. Boundary conditions are imposed by requiring that

$$f_{M+1} = 1 , \qquad g_{M+1} = 1 + \Delta , \qquad g_0 = 0 ,$$

and by inventing fictitious nodes at $x_{-1} = -\delta$, $x_{M+2} = (M + 2)\,\delta$ such that

$$g_{-1} = -g_1 , \qquad f_{-1} = f_1 , \qquad g_{M+2} = g_M + 2\delta \frac{\sin a}{\sin A} ,$$

$$f_{M+2} = f_M + \frac{2}{3}\delta \left\{ a \left(\cot a - \frac{\sin a \cos A}{\sin^2 A} \right) - \frac{\sin A}{\sin a} g''(1) \right\} .$$

We note that, by a standard limiting process, the condition $f'(0) = 0$ implies

$$\lim_{x \to 0} g''(x) = 0 .$$

Furthermore

$$f'(1) = \frac{a}{3} \left\{ \cot a - \frac{\sin a \cos A}{\sin^2 A} \right\} - \frac{\sin A}{3 \sin a} g''(1) .$$

The equation (18) is then discretised at the nodal points with the help of the boundary conditions and we obtain the following non-linear algebraic system

$$G_i(g_j, \lambda) = 0 , \qquad i, j = 1, 2, \ldots, M , \tag{20}$$

where we define $\lambda \equiv 1 + \Delta$. (Note the f_j can be eliminated by using (5).) We are interested in the solution space as the parameter λ varies. In order to traverse limit points (if any), we augment (20) by a scalar arc-length equation (this serves to define the arc-length)

$$N(\boldsymbol{g}, \lambda) = \sum_{j=1}^{M} \left(\frac{\partial g_j}{\partial s} \right)^2 + \left(\frac{\partial \lambda}{\partial s} \right)^2 - 1 = 0 , \tag{21}$$

where $\boldsymbol{g}$ is the solution vector of components $\{g_j\}$ and s is the arc-length along the solution curve.

Suppose we have a known solution point $\boldsymbol{p}_0 = (g^0, \lambda^0)$ at some value $s = s_0$ of the arc-length. We seek a solution $\boldsymbol{p}_1 = (\boldsymbol{g}^1, \lambda^1)$ at a nearby value of $s = s_1 = s_0 + \Delta s$. At point $\boldsymbol{p}_1$, (20—21) holds and to the leading order in Δs,

$$G_i(\boldsymbol{g}^0 + \dot{\boldsymbol{g}}^0 \Delta s, \lambda^0 + \dot{\lambda}^0 \Delta s) = G_i^0 + \left.\frac{\partial G_i}{\partial \boldsymbol{g}}\right|_0 \dot{\boldsymbol{g}} \Delta s + \left.\frac{\partial G_i}{\partial \lambda}\right|_0 \dot{\lambda} \Delta s = 0\,,$$

and a similar expression for the arc-length equation. In these equations the dot denotes a derivative with respect to s and the subscript (or superscript) '0' refers to an evaluation at $s = s^0$. Thus, the solution at $s = s_1$ may be obtained from $\boldsymbol{p}_0$ by the Euler's scheme:

$$\boldsymbol{g}^1 = \boldsymbol{g}^0 + \Delta \boldsymbol{g}\,, \qquad \lambda^1 = \lambda^0 + \Delta\lambda\,, \tag{22}$$

in which

$$G_{\boldsymbol{g}}^0 \Delta \boldsymbol{g} + G_\lambda^0 \Delta\lambda = -G^0\,, \qquad N_{\boldsymbol{g}}^0 \Delta \boldsymbol{g} + N_\lambda^0 \Delta\lambda = -N^0\,, \tag{23), (24}$$

where

$$\boldsymbol{G}^0 = \boldsymbol{G}(\boldsymbol{g}^0, \lambda^0)\,, \qquad N^0 = N(\boldsymbol{g}^0, \lambda^0)\,, \qquad G_{\boldsymbol{g}}^0 = \frac{\partial \boldsymbol{G}}{\partial \boldsymbol{g}}(\boldsymbol{g}^0, \lambda^0)\,,$$

$$G_\lambda^0 = \frac{\partial \boldsymbol{G}}{\partial \lambda}(\boldsymbol{g}^0, \lambda^0)\,, \qquad N_{\boldsymbol{g}}^0 = \frac{\partial N}{\partial \boldsymbol{g}}(\boldsymbol{g}^0, \lambda^0)\,, \qquad N_\lambda^0 = \frac{\partial N}{\partial \lambda}(\boldsymbol{g}^0, \lambda^0)\,.$$

We recall that a point $\boldsymbol{p} = (\boldsymbol{g}, \lambda)$ is a regular point if det $\boldsymbol{G}_{\boldsymbol{g}} \neq 0$. In the neighbourhood of a regular point, the Implicit Function Theorem guarantees the existence of a unique solution $\boldsymbol{g} = \boldsymbol{g}(\lambda)$. If det $\boldsymbol{G}_{\boldsymbol{g}} = 0$, then the point $\boldsymbol{p}$ is said to be *singular*; it can either be a limit point or a bifurcation point, refer to KELLER [4].

Now the system $\boldsymbol{Jx} = \boldsymbol{y}$ has a solution if and only if $\boldsymbol{y} \cdot \boldsymbol{n} = 0$, where $\boldsymbol{n}$ is a vector belonging to the null space of $\boldsymbol{J}$, say $\mathcal{N}(\boldsymbol{J}^\mathrm{T})$, where $\boldsymbol{J}^\mathrm{T}$ is the transpose of $\boldsymbol{J}$.

Let $\boldsymbol{n} \in \mathcal{N}(\boldsymbol{G}_{\boldsymbol{g}}^{0\mathrm{T}})$, and take the inner product of $\boldsymbol{n}$ with (23) to obtain

$$\boldsymbol{n} \cdot G_\lambda^0 \Delta\lambda = 0\,. \tag{25}$$

If G_λ^0 does not belong to the range of $G_{\boldsymbol{g}}^0$, then (25) implies that $\Delta\lambda = 0$ and we have a limit point. If G_λ^0 belongs to the range of $G_{\boldsymbol{g}}^0$, then it is orthogonal to $\boldsymbol{n}$ and $\Delta\lambda$ needs not be zero. In this case we have a bifurcation point.

Methods for negotiating bifurcating points are outlined in KELLER [4]. In this problem we find no bifurcation point and therefore a method due to KUBÍČEK and MAREK [3] is employed. In this method λ is defined to be g_{M+1} (this is the reason of our using the parameter $1 + \Delta$ rather than Δ itself), and g_k, k fixed, $1 < k < M + 1$, is treated as an independent variable. Provided that $k \neq M + 1$ near or at the limit point the resulting system matrix will not be singular and the system can be integrated across the limit point. Details of the method and the algorithm are given in KUBÍČEK and MAREK [3]. We find the method less comprehensive than KELLER's schemes [4], but very efficient in dealing with problems which have limit points.

Numerical results were obtained for a semi-cone angle $a = 45$ deg and $\varepsilon = 0, 0.3$. The case of $\varepsilon = 0$ corresponds to a neo-Hookean material. To verify the convergence of the numerical scheme, three different meshes were tried, with $M + 1 = 20$, 25 and 40. Limit points were obtained in all cases. At $\varepsilon = 0$ the limit point occurs at $1 + \Delta = 1.207$, and at $\varepsilon = 0.3$ the limit point occurs at $1 + \Delta = 1.238$. Richardson's extrapolation procedure has been used to find these limit points at zero mesh size, assuming that the algorithm converges quadratically. The solution diagrams are given on Figs. 2—3, where we plot C versus $\lambda \equiv 1 + \Delta$. The stable branches are in solid lines, and the unstable ones in broken lines. The solutions $g(x)$ at fixed values of λ are also given in Figs. 4—5. It can be seen that the unstable solution has a larger value of g in the neighbourhood of the centre line; this corresponds to a smaller radial displacement. Finally, although the limit points are of interest theoretically, we do not think that they are physical. They are most likely due to the artificial two-dimensionality of the problem, and may not occur in a finite deformation of a three-dimensional cone, with the free surface included. However, the present solution may be useful

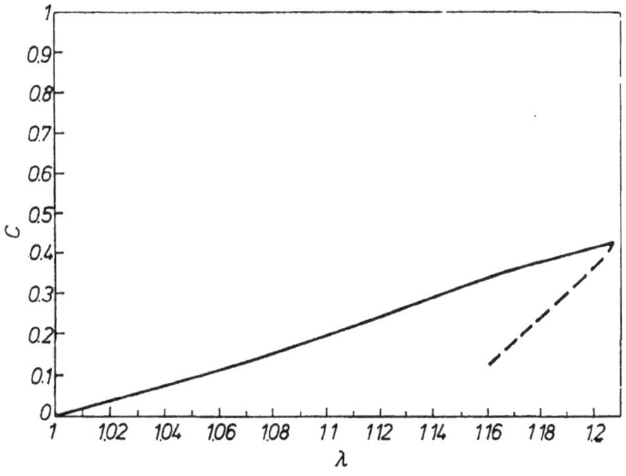

Fig. 2. The solution diagrams for $a = \pi/4$ and $\varepsilon = 0$.

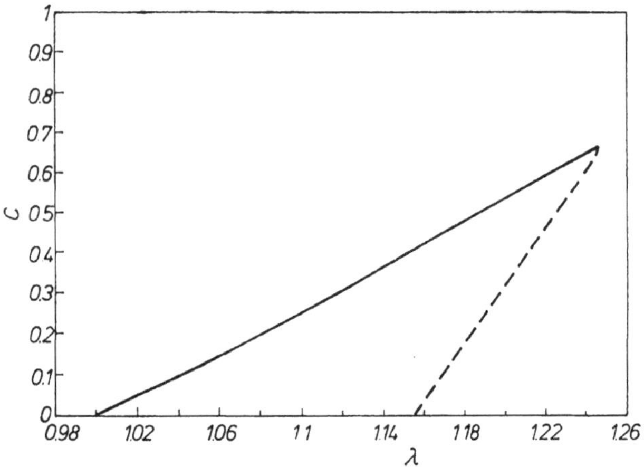

Fig. 3. The solution diagrams for $a = \pi/4$ and $\varepsilon = 0.3$

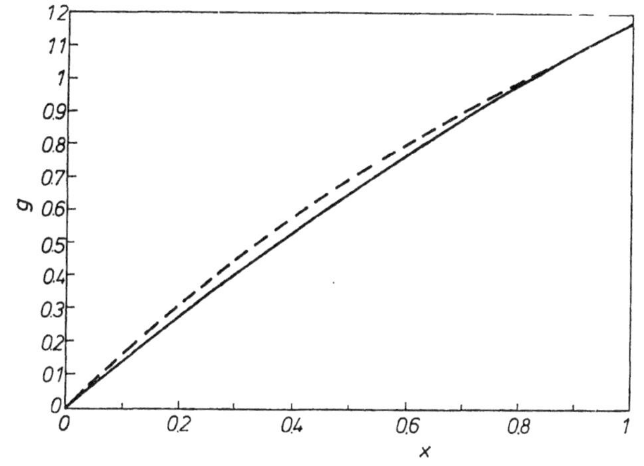

Fig. 4. Plots of $g(x)$ at $\lambda = 1.166$ for $\alpha = \pi/4$ and $\varepsilon = 0$. The stable solution is shown in solid line, the unstable one in broken line

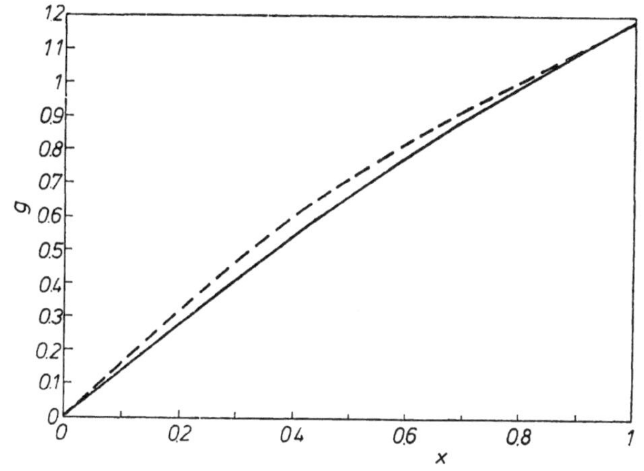

Fig. 5. Same as in Fig. 3, but with $\alpha = \pi/4$, $\varepsilon = 0.3$ and $\lambda = 1\,178$

as a bench mark solution for the computing analysts, who wish to test their general-purpose computer codes dealing with finite deformation problems. Because of the limit point at a critical value of $\lambda = \lambda_{\text{crit}}$, the finite deformation problem is no longer unique in the neighborhood of λ_{crit}; it is unstable in the sense of HADAMARD (see, for example [5]). Any numerical codes which are not designed to traverse limit points will cease to converge near a limit point.

Acknowledgement

N. PHAN-THIEN wishes to thank the Staff of the Department of Mechanical and Production Engineering for their marvelous hospitality.

References

1 PHAN-THIEN, N.; WALSH, W. P.: On the finite deformation of a two-dimensional elastic wedge. Z. angew. Math. Mech. 68 (1988) 9, 417–421.
2 MOONEY, M.: J. Appl. Phys. **11** (1940), 582–592.
3 KUBÍČEK, M.; MAREK, M.: Computational methods in bifurcation theory and dissipative structures. Springer-Verlag, Berlin 1983.
4 KELLER, H. B.: Applications of bifurcation theory. Academic Press, London 1977.
5 ERICKSEN, J. L.; TOUPIN, R. A.: J. Math. 8 (1956) 432–437.

Received August 3, 1987, in revised form March 23, 1988

Addresses: Dr. NHAN PHAN-THIEN, Reader in Mechanical Engineering, Department of Mechanical Engineering, The University of Sydney, Sydney, New South Wales, 2006 Australia; Dr. Y. T. CHEW, National University of Singapore, Mechanical, and Production Engineering Department, 10 Kent Ridge Crescent, Singapore 0511

BOOK REVIEWS

Bedford, T.; Swift, J. (eds.), New Directions in Dynamical Systems. Cambridge etc., Cambridge University Press 1988. XIII, 283 pp., £ 17.50. ISBN 0-521-34880-3 (London Mathematical Society Lecture Note Series 127)

Die nicht periodisch erscheinende mathematische Literatur konzentriert sich heute auf drei Schwerpunkte:

Da sind einmal die Lehrbücher, dann die Monographien und schließlich die Sammelbände von Forschungsarbeiten, die häufig anläßlich von Tagungen herausgegeben werden. Während es zwischen den beiden zuerst genannten Typen einen kontinuierlichen Übergang gibt, stehen Bücher der dritten Art separat und haben mehr den Charakter von Zeitschriftenbänden, die sich ausschließlich an Spezialisten wenden. Und doch wären mehr Veröffentlichungen, die in der Nähe des Gravizentrums der drei Typen liegen, wünschenswert. Ich meine damit Bücher, in denen mehrere erfolgreiche und bekannte Wissenschaftler unter einem gemeinsamen Hauptthema in Artikeln von etwa dreißig bis hundert Seiten über die Spezifika ihres engeren Gebietes berichten. Zu begrüßen sind die wenigen Erscheinungen dieser Art einmal weil sich hier auf verwandten Gebieten arbeitende Fachkollegen über die Umgebung ihres Arbeitsfeldes informieren können. Zum anderen liegt es nicht jedem, der sich in ein Gebiet einarbeiten will, ein Lehrbuch mit der durch Exaktheit und Vollständigkeit bei Definitionen und Beweisen bedingten Umständlichkeit durchzulesen, weil es ihm mehr hilft, schnell Probleme kennenzulernen und sich bei der Beschäftigung damit die notwendige Technik anzueignen.

Den Leser dieser Besprechung wird es nun nicht verwundern, wenn er hört, daß das vorliegende Buch zur soeben gepriesenen Gattung wissenschaftlicher Veröffentlichungen gehört. Nach einer kurzen Einleitung der Herausgeber enthält es sechs Artikel zu im Augenblick zentralen Themen der Dynamik. D. RAND gibt unter dem Titel „Universality and renormalization in dynamical systems" eine Einführung in die Probleme, die mit dieser bis heute in ihrer vollen Tragweite noch nicht abzuschätzenden faszinierenden Entdeckung zusammenhängen. In gewisser Weise verwandt hiermit ist das Thema der Arbeit „Smooth dynamics on the interval" von S. VAN STRIEN. Bifurkationen aus einer anderen Sicht werden in den Beiträgen von PAUL GLENDINNING „Global bifurcations in flows" und von I. STEWART „Bifurcations with symmetry" betrachtet. Ein zunächst vielleicht etwas abwegig erscheinendes Thema, das sich aber doch als inhaltsreich erwiesen hat, ist die Frage nach den Verknotungen von geschlossenen Orbits (d. h., von periodischen Lösungs- oder Bahnkurven) in dreidimensionalen Phasenräumen. Hierüber informiert P. HOLMES unter dem Titel „Knots and orbit genealogies in nonlinear oscillators". Schließlich geht N. G. LLOYD mit seinem Beitrag „Limit cycles of polynomial systems — some recent developments" auf eine klassische Frage ein, die zu den berühmten von Hilbert gestellten Problemen gehört.

Berlin H. G. BOTHE

Finite Elements in Fluids. Vol. 7. Chichester etc., John Wiley & Sons 1988. XI, 314 pp., £ 49.95. ISBN 0-471-91804-0 (Wiley Series in Numerical Methods in Engineering)

Das vorliegende Buch enthält 15 ausgewählte Beiträge, die als Vorträge anläßlich der Tagung: „Sixth International Conference on Finite Elements in Flow Problems", Antibes-Juan les Pins, Frankreich, 1986 gehalten wurden.

Neben den modernen Anwendungsformen der Finite-Elemente-Methode (FEM) sind darunter auch Beispiele zur Behandlung von Strömungsproblemen mittels Spektral- und Randintegralverfahren. Dies widerspiegelt die enge Verwandtschaft dieser Methoden und ist, nach Meinung der Herausgeber, auch repräsentativ für die gegenwärtigen Aktivitäten auf diesem Gebiet.

Eindeutiger Schwerpunkt der behandelten Probleme sind die kompressiblen Strömungen (6 Beiträge), gefolgt von Verbrennungsproblemen bzw. Strömungen mit chemischer Reaktion (3 Beiträge). Hier ergeben sich interessante Parallelen zu den Fällen, bei denen die Konvektion gegenüber der Diffusion dominiert, da die Behandlung chemischer Reaktionen zu ähnlichen numerischen Schwierigkeiten führen kann. In diesem Zusammenhang ist eine Tendenz zu beobachten, die zur Überwindung dieser Schwierigkeiten benutzten (Upwind-) Verfahren weiter zu „kultivieren".

Mit Hilfe einer mathematischen Begründung kann gezeigt werden, daß sogenannte „Streamline-Upwind/Petrov-Galerkin methods" (SUPG) neben ausreichender Stabilität eine gute Genauigkeit der Lösung garantieren.

Weiter behandelte Sachgebiete, die zum Teil auch übergreifend in einem Artikel behandelt werden, sind: — Strömungen mit freier Oberfläche, — Viscoelastische Probleme, — Inkompressible Strömungen, — Turbulenz.

Durch Weiterentwicklung der Computertechnik können immer komplexere Situationen numerisch erfaßt werden. Komplizierte 3-D-Simulationen mit sehr fein strukturierten Finite-Elemente-Netzen ($>$ 20000 Knoten) erfordern mehr als 10^{12} Gleitkommaoperationen und können nur von „Supercomputern" bewältigt werden.

Das Buch ist sorgfältig gestaltet und mit ausführlichen Literaturhinweisen am Schluß der Arbeiten versehen.

Berlin H. WILKE

Hagedorn, P.; Otterbein, S., Technische Schwingungslehre. Lineare Schwingungen diskreter mechanischer Systeme. Berlin etc., Springer-Verlag 1987. XII, 468 S., 184 Abb., DM 58,—. ISBN 3-540-18096-6

Fast gleichzeitig mit dem vorliegenden Buch erschien im gleichen Verlag die „Strukturdynamik" von GASCH und KNOTHE. Beide Bücher haben die gleiche Grundgliederung, unterscheiden sich jedoch in der Darstellung wesentlich. Die Tatsache zeigt jedoch, wie aktuell Schwingungsprobleme nach wie vor sind, und es ist deshalb besonders zu begrüßen, wenn durch unterschiedliche Betonungen die Vielfalt der Behandlungsmöglichkeit deutlich wird. Bei der „Technischen Schwingungslehre" von HAGEDORN und OTTERBEIN treten die Aspekte der technischen Anwendung gegenüber der Theorie zurück, und man spürt dort, wo sie angesprochen werden, eine Beschränkung auf allgemeine Grundbeispiele.

Die Gliederung entspricht der bewährten Form. Zunächst wird die Kinematik der Schwingungen im Zeit- und Frequenzbereich behandelt. Danach folgt eine ausführliche Darstellung der Schwingungen von Systemen mit einem Freiheitsgrad, wobei auch die nichtperiodische deterministische Erregung einen breiten Raum einnimmt. Es erweist sich immer als vorteilhaft, vor der Behandlung der Systeme mit endlich vielen Freiheitsgraden das „Minimalmodell" mit 2 Freiheitsgraden zu behandeln. Dabei ist es auch möglich, die Probleme gyroskopischer Effekte anzuschneiden. Mit beiden Abschnitten, die das halbe Buch füllen, ist die Grundlage für die Behandlung der Systeme mit endlich vielen Freiheitsgraden gelegt. In diesem Abschnitt fällt die relativ ausführliche Darstellung von Eigenwertabschätzungen sowie die schmale Behandlung des Eigenwertproblems auf. Hervorzuheben ist auch, daß die experimentelle Modalanalyse als Teilgebiet der Identifikation aufgegriffen wird, obwohl die Probleme, die sich bei der praktischen Realisierung der dargelegten Möglichkeiten ergeben, keine Erwähnung finden. Auch die Verfahren zur Aufstellung der Systemmatrizen werden, wenn man von denen zur Aufstellung der Bewegungsgleichungen absieht, nicht behandelt. Sehr zu begrüßen ist, daß in einem gesonderten Abschnitt die Anwendung der Fourier-Transformation in der Schwingungslehre dargestellt wird. Dabei spielen zwangsläufig die Zufallsschwingungen eine besondere Rolle.

Jeder Abschnitt des Buches enthält Übungsaufgaben. Daß dafür nur Lösungshinweise gegeben werden, wird sicher vom Lernenden anders als vom Lehrenden bewertet. Das Buch kann für ein Grundlagenstudium des Fachgebietes sehr empfohlen werden.

Dresden F. HOLZWEISSIG

ZAMM · Z. angew. Math. Mech. 69 (1989) 2, 73 – 81

STAAT, M., BALLMANN, J.

Zur Problematik tensorieller Verallgemeinerungen einachsiger nichtlinearer Materialgesetze

Durch Angabe von Gegenbeispielen wird gezeigt, daß es alleine aus der Kenntnis eines einachsigen Versuches nicht allgemein möglich ist, eindimensionale Stoffgesetze mehrdimensional zu verallgemeinern; insbesondere stößt man bei der nichtlinearen Erweiterung klassischer Moduln auf Widersprüche. Die in der Argumentation ergänzend benutzten dualen Materialgesetze ermöglichen wesentliche Vereinfachungen. Die zur vollständigen Identifikation der Stoffwerte benötigten Experimente werden besprochen.

It is shown by counterexamples that it is generally impossible to derive tensor constitutive equations from empirical scalar approximations of a single uniaxial experiment. Especially the use of generalized classical moduls leads to inconsistencies. Some examples are given, which show that the use of a dual constitutive equation allows considerable simplifications. The experiments necessary for a complete identification of material parameters are discussed.

При помощи противоречащих примеров доказывается невозможность выведения тензорных уравнений консистенции из эмпирических скалярных приближений одиночного эксперимента. Особенно применение обобщенных классических модулей водит к несовместимостям. Даются некоторые примеры показывающие, что применение двойственного уравнения консистенций позволяет значительные упрощения. Обсуждаются эксперыменты нужны для полной идентификации параметров материала.

1. Einleitung

Materialgesetze sind Approximationen realer Stoffe. Sie werden charakterisiert durch die Wahl geeigneter Funktionensysteme und einer Norm, in deren Sinne der Approximationsfehler minimiert werden soll. Stoffkennwerte sind zu bestimmende Koeffizienten der Approximation.

Geeignete Funktionensysteme lassen sich besonders leicht für skalare Funktionen einer unabhängigen Variablen finden. Daher überrascht es wenig, wenn, ausgehend von einachsigen Zug-Druck-Versuchen für linearelastische Werkstoffe, auch nichtlineare elastische sowie nichtelastische Stoffe im Ingenieurwesen weitgehend durch derartig gewonnene σ-ε-Diagramme charakterisiert werden. Und es werden viele Vorschläge veröffentlicht, wie diese Ergebnisse auf mehrdimensionale Spannungszustände zu übertragen seien [1, 2, 3].

Hier zeigt sich bereits eine erste Schwierigkeit des Begriffs vom einachsigen Stoffgesetz. Zumindest der sich einstellende Dehnungszustand ist dreiachsig. In der Praxis sind Versuche, die in Spannung und Dehnung einachsig sind, nicht realisierbar. Mechanische Stoffgesetze für Festkörper sind tensorwertige Funktionen (oder Funktionale), die Tensoren (z. B. Dehnungen) als Eingangsgrößen auf einen Tensor (z. B. Spannungen) als Ausgangsgröße abbilden. Im praktischen Experiment hat man daher z. B. die Komponenten der Eingangstensoren zu steuern und die Komponenten des Ausgangstensors zu beobachten.

Unabhängig von der Frage der Durchführbarkeit der benötigten Experimente kann die Kritik an einachsigen nichtlinearen Materialgesetzen bereits durch idealisierte, lediglich gedankliche Experimente begründet werden. Dazu wird angenommen, daß die Approximationsaufgabe für das Materialgesetz bereits gelöst und somit das Stoffgesetz als Tensorfunktion gegeben sei. Mit diesem Gesetz werden Experimente vorausberechnet. Dann wird untersucht, welche Experimente durchgeführt werden müßten, um die Stoffkennwerte zu rekonstruieren. Es zeigt sich, daß der einachsige Spannungszustand dazu nicht ausreicht. Eine andere Argumentation könnte auf der Tatsache aufbauen, daß sich ein bei unverformter, spannungsfreier Bezugskonfiguration isotroper Körper unter einem einachsigen Spannungszustand für eine weitere infinitesimale Dehnung lokal linear anisotrop verhält, wenn diese in Bezug auf die aktuelle Konfiguration gemessen wird.

Die Diskussion wird hier für homogene, isotrope, hyperelastische Materialien geführt, für die ein elastisches Potential existiert. Damit ist die Zahl möglicher Funktionensysteme zwar von vornherein eingeschränkt, doch läßt sich die Argumentation leicht auf andere (insbesondere isotrope) Tensorfunktionen übertragen. Für kompressible Stoffe werden duale Materialgesetze genutzt, die sich aus dem komplementären elastischen Potential ableiten lassen. Die Ergebnisse lassen sich wesentlich klarer herausarbeiten, wenn man, anders als üblich, beide Materialgesetze gleichberechtigt benutzt.

2. Isotropes, kompressibles, elastisches Material

2.1 Approximation realer Stoffe

Das Stoffgesetz eines isotropen Punktkontinuums kann allgemein durch eine isotrope Tensorfunktion der Form

$$\boldsymbol{Y}(\boldsymbol{X}) = \varphi_0 \boldsymbol{1} + \varphi_1 \boldsymbol{X} + \varphi_2 \boldsymbol{X}^2 \tag{2.1a}$$

dargestellt werden. Darin sind $\boldsymbol{Y}$ und $\boldsymbol{X}$ Tensoren 2. Stufe; $\boldsymbol{1}$ ist der Einheitstensor. Für den Beweis dieser Darstellung benutzt man vorteilhaft das Theorem von Cayley-Hamilton, mit dessen Hilfe man (2.1a) auch auf drei andere Potenzen des Tensors $\boldsymbol{X}$ umschreiben kann. Existiert die Inverse $\boldsymbol{X}^{-1}$, so lautet eine äquivalente Darstellung

$$\boldsymbol{Y}(\boldsymbol{X}) = \psi_0 \boldsymbol{1} + \psi_1 \boldsymbol{X} + \psi_{-1} \boldsymbol{X}^{-1} \,. \tag{2.1b}$$

Darin sind die Koeffizienten $\varphi_0, \varphi_1, \varphi_2, \psi_0, \psi_1$ und ψ_{-1} skalare Funktionen der Invarianten von $\boldsymbol{X}$ und als solche selbst Invarianten. Da die Funktionen i. a. voneinander unabhängig sind, können sie auch nicht auf eine Funktion (etwa auf den nichtlinear erweiterten Youngschen Elastizitätsmodul) zurückgeführt werden. Nur wenn man die Dimension des Problems reduziert, läßt sich die Anzahl der Materialfunktionen allgemein verringern. Dann gilt (2.1) jedoch nur für Tensoren in einem Raum geringerer Dimension. Es ist zu untersuchen, ob eine durchgeführte Reduktion in eindeutiger Weise zurücknehmbar ist.

Bei Rand- und Anfangsrandwertproblemen setzt man üblicherweise ein geeignetes objektives Dehnungsmaß als $\boldsymbol{X}$ ein und erhält als $\boldsymbol{Y}$ den zugehörigen Spannungstensor. Im Experiment wird umgekehrt i. a. die Spannung als Argumenttensor $\boldsymbol{X}$ angesetzt. Denn selbst im sogenannten dehnungsgesteuerten einachsigen Zug-Druckversuch wird die Querkontraktion nicht gesteuert, sondern de facto die Spannungsrandbedingung auf den freien Rändern.

Hier seien der 2. Piola-Kirchhoff-Spannungstensor $\boldsymbol{\sigma}$ und der Greensche Verzerrungstensor $\boldsymbol{G}$

$$\boldsymbol{G} = \tfrac{1}{2}(\boldsymbol{C} - \boldsymbol{1}) = \tfrac{1}{2}(\boldsymbol{H} + \boldsymbol{H}^{\mathrm{T}} + \boldsymbol{H}^{\mathrm{T}}\boldsymbol{H}) \tag{2.2}$$

gewählt. Dabei ist $\boldsymbol{H}$ der materiell gebildete Verschiebungsgradient, $\boldsymbol{C} = \boldsymbol{F}^{\mathrm{T}}\boldsymbol{F}$ der rechte Cauchy-Green-Tensor und $\boldsymbol{F}$ der materiell gebildete Deformationsgradient, der die als spannungsfrei angenommene Bezugskonfiguration auf die aktuelle Konfiguration abbildet. Der Verzerrungstensor (hier $\boldsymbol{G}$) berücksichtigt die geometrische Nichtlinearität, während die physikalische Nichtlinearität von der Form der Koeffizienten $\varphi_0, \ldots, \psi_{-1}$ bestimmt wird. In der Bezugskonfiguration sind $\boldsymbol{G}$ und $\boldsymbol{\sigma}$ Nulltensoren $\boldsymbol{0}$, weshalb die Form (2.1a) benutzt wird. Wegen der Isotropie fallen die Hauptachsen $\boldsymbol{e}_{\mathrm{I}}, \boldsymbol{e}_{\mathrm{II}}, \boldsymbol{e}_{\mathrm{III}}$ von $\boldsymbol{G}$ und $\boldsymbol{\sigma}$ zusammen, so daß

$$\boldsymbol{G} = G_{\mathrm{I}}\boldsymbol{e}_{\mathrm{I}} \circ \boldsymbol{e}_{\mathrm{I}} + G_{\mathrm{II}}\boldsymbol{e}_{\mathrm{II}} \circ \boldsymbol{e}_{\mathrm{II}} + G_{\mathrm{III}}\boldsymbol{e}_{\mathrm{III}} \circ \boldsymbol{e}_{\mathrm{III}}\,, \qquad \boldsymbol{\sigma} = \sigma_{\mathrm{I}}\boldsymbol{e}_{\mathrm{I}} \circ \boldsymbol{e}_{\mathrm{I}} + \sigma_{\mathrm{II}}\boldsymbol{e}_{\mathrm{II}} \circ \boldsymbol{e}_{\mathrm{II}} + \sigma_{\mathrm{III}}\boldsymbol{e}_{\mathrm{III}} \circ \boldsymbol{e}_{\mathrm{III}}\,. \tag{2.3 a, b}$$

Mit den linearen, quadratischen und kubischen Hauptinvarianten

$$\left.\begin{aligned} &I_1 := \operatorname{Sp}\boldsymbol{G} = G_{\mathrm{I}} + G_{\mathrm{II}} + G_{\mathrm{III}}\,, \quad I_2 := \tfrac{1}{2}(\operatorname{Sp}\boldsymbol{G}^2 - I_1^2) = -G_{\mathrm{I}}G_{\mathrm{II}} - G_{\mathrm{I}}G_{\mathrm{III}} - G_{\mathrm{II}}G_{\mathrm{III}}\,,\\ &I_3 := \det\boldsymbol{G} = G_{\mathrm{I}}G_{\mathrm{II}}G_{\mathrm{III}}\,, \quad J_1 := \operatorname{Sp}\boldsymbol{\sigma} = \sigma_{\mathrm{I}} + \sigma_{\mathrm{II}} + \sigma_{\mathrm{III}}\,,\\ &J_2 := \tfrac{1}{2}(\operatorname{Sp}\boldsymbol{\sigma}^2 - J_1^2) = -\sigma_{\mathrm{I}}\sigma_{\mathrm{II}} - \sigma_{\mathrm{I}}\sigma_{\mathrm{III}} - \sigma_{\mathrm{II}}\sigma_{\mathrm{III}}\,, \quad J_3 := \det\boldsymbol{\sigma} = \sigma_{\mathrm{I}}\sigma_{\mathrm{II}}\sigma_{\mathrm{III}} \end{aligned}\right\} \tag{2.4 a, ..., f}$$

lautet das Stoffgesetz für isotropes, kompressibles, elastisches Material

$$\boldsymbol{\sigma} = \tau_{0(I_1,I_2,I_3)}\boldsymbol{1} + \tau_{1(I_1,I_2,I_3)}\boldsymbol{G} + \tau_{2(I_1,I_2,I_3)}\boldsymbol{G}^2\,, \qquad \boldsymbol{G} = \varepsilon_{0(J_1,J_2,J_3)}\boldsymbol{1} + \varepsilon_{1(J_1,J_2,J_3)}\boldsymbol{\sigma} + \varepsilon_{2(J_1,J_2,J_3)}\boldsymbol{\sigma}^2\,. \tag{2.5 a, b}$$

Für das DE SAINT-VENANT-KIRCHHOFF-Material gilt unter Einbeziehung geometrischer Nichtlinearitäten das lineare Materialgesetz

$$\boldsymbol{\sigma} = \lambda I_1\boldsymbol{1} + 2\mu\boldsymbol{G}; \tag{2.6a}$$

λ und μ sind die Lamé-Koeffizienten der linearen Approximation. Das duale Stoffgesetz ist die Umkehrfunktion:

$$\boldsymbol{G} = \Lambda J_1\boldsymbol{1} + 2\Gamma\boldsymbol{\sigma} \tag{2.6b}$$

mit

$$\Lambda = \frac{-\lambda}{2\mu(3\lambda + 2\mu)}\,, \qquad \Gamma = \frac{1}{2\mu}\,. \tag{2.7 a, b}$$

Für das lineare Stoffgesetz (2.6a) existiert eindeutig ein spezifisches elastisches Potential W (ϱ Dichte in der Bezugskonfiguration):

$$\varrho W = \varrho U_{(I_1,I_2)} = \frac{\lambda + 2\mu}{2}I_1^2 + 2\mu I_2\,. \tag{2.8a}$$

Materialien, für die sich das Stoffgesetz durch Variation eines solchen Potentials ableiten läßt, nennt man Green-elastisch oder hyperelastisch. Hängt das Potential ganz allgemein auch von der kubischen Invarianten ab — also $W = U(I_1, I_2, I_3)$ — dann erhält man

$$\boldsymbol{\sigma} = \varrho\frac{\partial W}{\partial \boldsymbol{G}}\,, \qquad \boldsymbol{\sigma} = \varrho\left(\frac{\partial U}{\partial I_1} - I_1\frac{\partial U}{\partial I_2} - I_2\frac{\partial U}{\partial I_3}\right)\boldsymbol{1} + \varrho\left(\frac{\partial U}{\partial I_2} - I_1\frac{\partial U}{\partial I_3}\right)\boldsymbol{G} + \varrho\left(\frac{\partial U}{\partial I_3}\right)\boldsymbol{G}^2\,. \tag{2.9 a, b}$$

Durch Vergleich mit (2.5a) identifiziert man τ_0, τ_1 und τ_2. Diese skalaren Funktionen müssen notwendige und hinreichende Integrabilitätsbedingungen in Form dreier partieller Differentialgleichungen erfüllen, damit das Potential existiert [4]. Das ist für lineare Stoffgesetze der Fall.

Eine Legendre-Transformation

$$\varrho U_{\mathrm{c}}(J_1, J_2, J_3) = \boldsymbol{\sigma} : \boldsymbol{G} - \varrho U(I_1, I_2, I_3) \tag{2.10}$$

liefert das komplementäre (konjugierte) spezifische elastische Potential $W_{\mathrm{c}} = U_{\mathrm{c}}(J_1, J_2, J_3)$. Die doppelte Überschiebung $\boldsymbol{\sigma} : \boldsymbol{G} = \operatorname{Sp}(\boldsymbol{\sigma}\boldsymbol{G})$,

$$\boldsymbol{\sigma} : \boldsymbol{G} = \tau_0 \operatorname{Sp}\boldsymbol{G} + \tau_1 \operatorname{Sp}\boldsymbol{G}^2 + \tau_2 \operatorname{Sp}\boldsymbol{G}^3 \quad \text{oder} \quad \boldsymbol{\sigma} : \boldsymbol{G} = \varepsilon_0 \operatorname{Sp}\boldsymbol{\sigma} + \varepsilon_1 \operatorname{Sp}\boldsymbol{\sigma}^2 + \varepsilon_2 \operatorname{Sp}\boldsymbol{\sigma}^3\,, \tag{2.11 a, b}$$

erlaubt prinzipiell den Übergang von den Invarianten von $\boldsymbol{G}$ zu jenen von $\boldsymbol{\sigma}$ als neue Variablen. Im linearen Fall nach (2.6a) ist $\boldsymbol{\sigma} : \boldsymbol{G} = 2\varrho W_{\mathrm{c}}$. Ferner gelten die Beziehungen

$$\frac{\lambda + 2\mu}{2}I_1^2 = \frac{\lambda + 2\mu}{2(3\lambda + 2\mu)^2}J_1^2\,, \qquad 2\mu I_2 = \frac{(3\lambda + 4\mu)\lambda}{2\mu(3\lambda + 2\mu)}J_1^2 + \frac{1}{2\mu}J_2\,. \tag{2.12 a, b}$$

Mit (2.7) hat man dann

$$\varrho W_c = \varrho U_c(J_1, J_2) = \frac{\Lambda + 2\Gamma}{2} J_1^2 + 2\Gamma J_2 \tag{2.8b}$$

und für das Stoffgesetz (2.6b)

$$G = \varrho \frac{\partial W_c}{\partial \sigma}. \tag{2.9c}$$

Gleichung (2.9c) ist analog (2.9b) auszuführen. Die Koeffizienten ε_0, ε_1 und ε_2 erfüllen analoge Integrabilitätsbedingungen wie die Koeffizienten τ_0, τ_1 und τ_2. Das Konzept des elastischen Potentials erlaubt die Darstellung des Stoffgesetzes in Form einer skalaren Funktion im dreidimensionalen Raum der Invarianten von $\boldsymbol{G}$ bzw. $\boldsymbol{\sigma}$. Man kann also ein linear unabhängiges Funktionensystem geeignet wählen, um damit ein Materialgesetz (2.9) zu konstruieren. Die Stoffkennwerte sind dann etwa durch eine beste Approximation gemessener Potentiale W und W_c (im Sinne kleinster Fehlerquadrate oder auch einer anderen Norm) zu bestimmen. So gesehen stellt das lineare Stoffgesetz (2.8) eine Approximation durch Polynome dar, wobei die Eigenwerte von $\boldsymbol{\sigma}$ und $\boldsymbol{G}$ quadratisch vorkommen.

Die vollständigen, in den Eigenwerten kubischen Polynome lauten

$$\varrho W = \varrho U(I_1, I_2, I_3) = \frac{\lambda + 2\mu}{2} I_1^2 + 2\mu I_2 + \frac{l + 2m}{3} I_1^3 + 2m I_1 I_2 + n I_3,$$

$$\varrho W_c = \varrho U_c(J_1, J_2, J_3) = \frac{\Lambda + 2\Gamma}{2} J_1^2 + 2\Gamma J_2 + \frac{L + 2M}{3} J_1^3 + 2M J_1 J_2 + N J_3, \tag{2.13 a, b}$$

wobei alle Kennwerte zahlenmäßig neu zu bestimmen sind. In ihrer Abhängigkeit von der Approximation stellen sie offenbar keine Stoffkonstanten dar. Ein Kennzeichen für physikalische Nichtlinearität ist in diesem Zusammenhang i. a. $W_c \neq W$. Durch Differenzieren von (2.13 a, b) erhält man das Stoffgesetz für das sog. Murnaghan-Material

$$\left.\begin{aligned} \boldsymbol{\sigma} &= [\lambda I_1 + l I_1^2 + (2m - n) I_2] \boldsymbol{1} + [2\mu + (2m - n) I_1] \boldsymbol{G} + n \boldsymbol{G}^2, \\ \boldsymbol{G} &= [\Lambda J_1 + L J_1^2 + (2M - N) J_2] \boldsymbol{1} + [2\Gamma + (2M - N) J_1] \boldsymbol{\sigma} + N \boldsymbol{\sigma}^2. \end{aligned}\right\} \tag{2.14 a, b}$$

Beschränkt man sich nicht auf hyperelastische Stoffe, so ist in Gleichung (2.14) ein weiterer Koeffizient zu bestimmen [5]. Die experimentelle Aufgabe ist hier also vereinfacht worden.

Jede dieser Darstellungen (2.13 a, b) und (2.14 a, b) approximiert das reale Material mit der vorgegebenen Fehlerordnung. Charakteristisch für die nichtlinearen Gesetze ist aber, daß sich die Invarianten von $\boldsymbol{G}$ nicht, wie im linearen Fall in Gleichung (2.12) geschehen, durch die Invarianten von $\boldsymbol{\sigma}$ ausdrücken lassen. Man wird auf ein System transzendenter Gleichungen geführt, das sich analytisch nicht lösen läßt [4]. Deshalb sind (2.14a) und (2.14b) zwei nicht ineinander überführbare Approximationen eines realen Materials, deren Approximationsfehler von gleicher Ordnung quadratisch in den Eigenwerten ist. Der Fehler ist im Sinne derselben Norm zu minimieren. So ist auch der Vorschlag in [6] zu verstehen, die Umrechnung der Invarianten beider Stoffgesetze lediglich mit der Genauigkeit der Murnaghan-Approximation durchzuführen. Setzt man (2.14a) und

$$J_1 \approx (3\lambda + 2\mu) I_1 + (3l + 2m I_1^2) + (6m - n) I_2, \qquad J_2 \approx -(3\lambda + 4\mu) \lambda I_1^2 + 4\mu^2 I_2 \tag{2.12 c, d}$$

in (2.14 b) ein und vernachlässigt alle in den Eigenwerten von $\boldsymbol{G}$ kubischen Terme, so findet man durch Koeffizientenvergleich neben (2.7 a, b)

$$L \approx \frac{-l}{(3\lambda + 2\mu)^3} + \frac{\lambda n(3\lambda^2 + 6\lambda\mu + 4\mu^2)}{8\mu^3(3\lambda + 2\mu)^3}, \qquad M \approx \frac{-\lambda n - 4\mu m}{16\mu^3(3\lambda + 2\mu)}, \qquad N \approx \frac{-n}{8\mu^3}. \tag{2.7 c, d, e}$$

Die Umrechnung in entgegengesetzter Richtung erhält man durch Vertauschen von λ, μ, l, m und n mit Λ, Γ, L, M und N in den Gleichungen (2.7 a–e).

Es muß aber betont werden, daß die Gleichungen (2.12 c, d) nur für hinreichend kleine $\boldsymbol{G}$ bzw. $\boldsymbol{\sigma}$ konvergente oder genügend schnell konvergente Entwicklungen darstellen. Selbst in (2.7 a, b) ist das Gleichheitszeichen bei der Approximation tatsächlich durchgeführter Experimente in der Regel nicht erreichbar. Deshalb wird in dieser Arbeit vorgeschlagen, keine der beiden Approximationen (2.14) als primär gegeben anzusehen und beide getrennt durchzuführen. Dabei müssen, unabhängig voneinander, sowohl $\boldsymbol{G}$ als auch $\boldsymbol{\sigma}$ als fehlerbehaftet betrachtet werden. Die Materialgesetze (2.14) sind dann nicht mehr mit (2.7 a–e) ineinander überführbar. Des weiteren bedeutet der Vorschlag, daß in der Praxis zu jedem nichtlinearelastischen Material die Koeffizienten zweier Approximationen zu ermitteln sind. Solche Daten liegen bisher für das duale Gesetz nicht vor. Deshalb werden geeignete Experimente zur Bestimmung der Stoffwerte diskutiert. Das duale Materialgesetz (2.14b) erweist sich dabei als leichter handhabbar. Denkt man den hier gemachten Vorschlag weiter, so wird man auch zulassen, daß in (2.5a) und (2.5b) zwei Ansätze unabhängig voneinander geeignet gewählt werden.

2.2 Experimente mit einfachen Dehnungen

Materialkennwerte für schwach nichtlineare Stoffe werden oft durch Ultraschallaufzeitmessungen in dynamischen Experimenten ermittelt [7]. Die Verzerrungszustände sind dabei inhomogen. Für stärkere Nichtlinearitäten und mit Blick auf die Zielsetzung, die Unzulänglichkeiten einachsiger Versuche zu beleuchten, sollen hier statische Experimente gedanklich durchgeführt werden. Statisch können homogene Verzerrungszustände benutzt werden, um

Stoffparameter zu bestimmen. Denn homogene Verzerrungszustände sind in jedem elastischen Körper allein durch Oberflächenkräfte steuerbar [5, 8]; sie lassen sich vollständig berechnen, ohne auf Randwertprobleme zu führen, die erst nach Kenntnis des Stoffgesetzes lösbar wären.

Jeder Teil eines homogen verzerrten homogenen isotropen Körpers steht stellvertretend für das Ganze, weil in jedem Punkt derselbe Zustand herrscht; kurz grad $\boldsymbol{F} \equiv \boldsymbol{0}$. Spezielle homogene Zustände sind einfache Dehnungen, für die gilt

$$\boldsymbol{G} = [\boldsymbol{e}_{\rm I} \circ \boldsymbol{e}_{\rm I} + a(\boldsymbol{e}_{\rm II} \circ \boldsymbol{e}_{\rm II} + \boldsymbol{e}_{\rm III} \circ \boldsymbol{e}_{\rm III})]\, G_{\rm I} \quad \text{oder} \quad \boldsymbol{G} = [\boldsymbol{e}_{\rm I} \circ \boldsymbol{e}_{\rm I} + \boldsymbol{e}_{\rm II} \circ \boldsymbol{e}_{\rm II} + b\boldsymbol{e}_{\rm III} \circ \boldsymbol{e}_{\rm III}]\, G_{\rm I}\,. \tag{2.15 a, b}$$

Wegen Isotropie gilt dann für die korrespondierenden Spannungszustände

$$\boldsymbol{\sigma} = [\boldsymbol{e}_{\rm I} \circ \boldsymbol{e}_{\rm I} + \alpha(\boldsymbol{e}_{\rm II} \circ \boldsymbol{e}_{\rm II} + \boldsymbol{e}_{\rm III} \circ \boldsymbol{e}_{\rm III})]\, \sigma_{\rm I} \quad \text{oder} \quad \boldsymbol{\sigma} = [\boldsymbol{e}_{\rm I} \circ \boldsymbol{e}_{\rm I} + \boldsymbol{e}_{\rm II} \circ \boldsymbol{e}_{\rm II} + \beta\boldsymbol{e}_{\rm III} \circ \boldsymbol{e}_{\rm III}]\, \sigma_{\rm I}\,. \tag{2.16 a, b}$$

An Hand der einfachen Dehnung sollen die Gleichungen (2.14) diskutiert werden. Dabei ist zu beachten, daß meistens α und β im Experiment vorgegeben werden, so daß sich a und b nicht steuerbar gemäß dem Stoffgesetz einstellen. Die zur vollständigen Identifikation der Materialkennwerte nutzbaren Experimente werden zur leichteren Bezugnahme zusammengefaßt bereitgestellt.

i) *Verzerrungen bei spannungsgesteuerten Experimenten*

i) a) Experimente gemäß Gleichungen (2.16a):

Mit $J_1 = (1 + 2\alpha)\, \sigma_{\rm I}$ und $J_2 = -(\alpha + 2)\, \alpha\sigma_{\rm I}^2$ folgt aus Gleichung (2.14b)

$$\left.\begin{aligned} G_{\rm I} &= \{\Lambda(1 + 2\alpha) + 2\Gamma + [L(1 + 2\alpha)^2 + 2M(1 - \alpha^2) + N\alpha^2]\, \sigma_{\rm I}\}\, \sigma_{\rm I}\,, \\ G_{\rm II} = G_{\rm III} &= \{\Lambda(1 + 2\alpha) + 2\alpha\Gamma + [L(1 + 2\alpha)^2 + 2M\alpha(\alpha - 1) + N\alpha]\, \sigma_{\rm I}\}\, \sigma_{\rm I}\,. \end{aligned}\right\} \tag{2.17 a, b}$$

Spezialfälle: hydrostatischer Spannungszustand ($\alpha = 1$):

$$G_{\rm I} = G_{\rm II} = G_{\rm III} = [3\Lambda + 2\Gamma + (9L + N)\, \sigma_{\rm I}]\, \sigma_{\rm I}; \tag{2.18}$$

einachsiger Spannungszustand ($\alpha = 0$):

$$G_{\rm I} = [\Lambda + 2\Gamma + (L + 2M)\, \sigma_{\rm I}]\, \sigma_{\rm I}\,, \qquad G_{\rm II} = G_{\rm III} = [\Lambda + L\sigma_{\rm I}]\, \sigma_{\rm I}\,. \tag{2.19 a, b}$$

i) b) Experimente gemäß Gleichung (2.16b):

Mit $J_1 = (2 + \beta)\, \sigma_{\rm I}$ und $J_2 = -(1 + 2\beta)\, \sigma_{\rm I}^2$ folgt aus Gleichung (2.14b)

$$\left.\begin{aligned} G_{\rm I} = G_{\rm II} &= \{\Lambda(2 + \beta) + 2\Gamma + [L(2 + \beta)^2 + 2M(1 - \beta) + N\beta]\, \sigma_{\rm I}\}\, \sigma_{\rm I}\,, \\ G_{\rm III} &= \{\Lambda(2 + \beta) + 2\beta\Gamma + [L(2 + \beta)^2 + 2M(\beta^2 - 1) + N]\, \sigma_{\rm I}\}\, \sigma_{\rm I}\,. \end{aligned}\right\} \tag{2.20 a, b}$$

Spezialfälle: hydrostatischer Spannungszustand ($\beta = 1$): Gleichung (2.18);

gleiche zweiachsige Spannung ($\beta = 0$):

$$G_{\rm I} = G_{\rm II} = [2(\Lambda + \Gamma) + (4L + 2M)\, \sigma_{\rm I}]\, \sigma_{\rm I}\,, \qquad G_{\rm III} = [2\Lambda + (4L - 2M + N)\, \sigma_{\rm I}]\, \sigma_{\rm I}\,. \tag{2.21 a, b}$$

ii) *Spannungen bei verzerrungsgesteuerten Experimenten*

ii) a) Experimente gemäß Gleichung (2.15 a)

Mit $I_1 = (1 + 2a)\, G_{\rm I}$ und $I_2 = -(a + 2)\, aG_{\rm I}^2$ folgt aus Gleichung (2.14a) je eine Gleichung für $\sigma_{\rm I}$ und eine für $\sigma_{\rm II} = \sigma_{\rm III}$. Multiplizieren der ersten Gleichung mit α gemäß Beziehung (2.16a) und Gleichsetzen mit der zweiten liefert unter der Annahme $G_{\rm I} \neq 0$ eine quadratische Gleichung für a ($G_{\rm I} = 0$ ist trivial). Ihre Lösung ist

$$G_{\rm II} = G_{\rm III} = aG_{\rm I} = \frac{2\lambda(1 - \alpha) + 2\mu + [4l(1 - \alpha) - 2m + n]\, G_{\rm I}}{2[4l(1 - \alpha) + 2m(1 + \alpha) - n\alpha]} \times$$

$$\times \left\{-1 + S\sqrt{1 - \frac{4[4l(1 - \alpha) + 2m(1 + \alpha) - n\alpha]\, \{\lambda(1 - \alpha) - 2\mu\alpha + [l(1 - \alpha) - 2m\alpha]\, G_{\rm I}\}\, G_{\rm I}}{\{2\lambda(1 - \alpha) + 2\mu + [4l(1 - \alpha) - 2m + n]\, G_{\rm I}\}^2}}\right\},$$

$$\sigma_{\rm I} = \{\lambda(1 + 2a) + 2\mu + [l(1 + 2a)^2 + 2m(1 - a^2) + na^2]\, G_{\rm I}\}\, G_{\rm I}\,,$$

mit

$$S := \operatorname{sign}\left(\frac{\lambda(1 - \alpha) + \mu}{4l(1 - \alpha) + 2m(1 + \alpha) - n\alpha}\right). \tag{2.22 a, b, c}$$

Hier, wie im folgenden, wurde jeweils das Vorzeichen der Wurzel gewählt, das bei verschwindenden Dehnungen zu Spannungsfreiheit führt.

Spezialfälle: hydrostatischer Spannungszustand ($\alpha = 1$): Es folgt $G_{\rm II} = G_{\rm III} = G_{\rm I}$ bzw. $a = 1$ und

$$\sigma_{\rm I} = \sigma_{\rm II} = \sigma_{\rm III} = [3\lambda + 2\mu + (9l + n)\, G_{\rm I}]\, G_{\rm I}; \tag{2.23}$$

einachsiger Spannungszustand ($\alpha = 0$): Mit $a_0 := a(\alpha = 0)$ folgt

$$G_{\rm II} = G_{\rm III} = a_0 G_{\rm I} = \frac{2\lambda + 2\mu + (4l - 2m + n)\, G_{\rm I}}{2(4l + 2m)} \left\{-1 + S\sqrt{1 - \frac{4(4l + 2m)\, (\lambda + lG_{\rm I})\, G_{\rm I}}{[2\lambda + 2\mu + (4l - 2m + n)\, G_{\rm I}]^2}}\right\},$$

$$\sigma_{\rm I} = (\lambda + 2\mu + 4lG_{\rm II})\, G_{\rm I} + (l + 2m)\, G_{\rm I}^2 + 2\lambda G_{\rm II} + (4l - 2m + n)\, G_{\rm II}^2\,, \tag{2.24 a, b}$$

mit

$$S := \operatorname{sign}\left(\frac{\lambda + \mu}{2(l + m)}\right). \tag{2.24 c}$$

ii) b) Experimente gemäß Gleichung (2.15b):

Mit $I_1 = (2 + b) G_\mathrm{I}$ und $I_2 = -(1 + 2b) G_\mathrm{I}^2$ folgt aus Gleichung (2.14a) ähnlich wie oben

$$G_\mathrm{III} = bG_\mathrm{II} = bG_\mathrm{I} = \frac{\lambda(1-\beta) + 2\mu + [4l(1-\beta) + 2m\beta + n\beta]\, G_\mathrm{I}}{2[l(1-\beta) + 2m]} \times$$

$$\times \left\{-1 + S\sqrt{1 - \frac{4[l(1-\beta) + 2m]\,\{2\lambda(1-\beta) - 2\mu\beta + [4l(1-\beta) - 2m(1+\beta) + n(1-2\beta)]\, G_\mathrm{I}\}\, G_\mathrm{I}}{\{\lambda(1-\beta) + 2\mu + [4l(1-\beta) + 2m\beta + n\beta]\, G_\mathrm{I}\}^2}}\right\},$$

$$\sigma_\mathrm{I} = \{\lambda(2 + b) + 2\mu + [l(2 + b)^2 + 2m(1 - b) + n(2 - b)]\, G_\mathrm{I}\}\, G_\mathrm{I},$$

mit

$$S := \operatorname{sign}\left(\frac{\lambda(1-\beta) + 2\mu}{l(1-\beta) + 2m}\right). \tag{2.25 a, b, c}$$

Spezialfälle: hydrostatischer Spannungszustand ($\beta = 1$): Gleichung (2.23);
gleiche zweiachsige Spannung ($\beta = 0$): Mit $b_0 := b$ ($\beta = 0$) folgt

$$G_\mathrm{III} = b_0 G_\mathrm{II} = b_0 G_\mathrm{I} = \frac{\lambda + 2\mu + 4lG_\mathrm{I}}{2l + 4m}\left\{-1 + S\sqrt{1 - \frac{4(l + 2m)\,\{2\lambda + [4l - 2m + n]\, G_\mathrm{I}\}\, G_\mathrm{I}}{[\lambda + 2\mu + 4lG_\mathrm{I}]^2}}\right\},$$

$$\sigma_\mathrm{I} = (2\lambda + 2\mu)\, G_\mathrm{I} + (4l + 2m + 2n)\, G_\mathrm{I}^2 + (4l - 2m - n)\, G_\mathrm{I} G_\mathrm{III} + \lambda G_\mathrm{III} + lG_\mathrm{III}^2,$$

mit

$$S := \operatorname{sign}\left(\frac{\lambda + 2\mu}{l + 2m}\right). \tag{2.26 a, b, c}$$

2.3 Bestimmbarkeit der Materialkennwerte

Für viele elastische Stoffe kann der einachsige Spannungszustand $\sigma = \sigma_\mathrm{I} \boldsymbol{e}_\mathrm{I} \circ \boldsymbol{e}_\mathrm{I}$ genügend genau erzeugt werden, um den linearen Youngschen Elastizitätsmodul E und die Poissonsche Querkontraktionszahl ν

$$E = \frac{1}{\Lambda + 2\Gamma} = \frac{\mu(3\lambda + 2\mu)}{\lambda + \mu}, \qquad \nu = \frac{-\Lambda}{\Lambda + 2\Gamma} = \frac{\lambda}{2(\lambda + \mu)} \tag{2.27), (2.28}$$

zu bestimmen. Dazu wird der Dehnungszustand

$$\boldsymbol{G} = G_\mathrm{I}[\boldsymbol{e}_\mathrm{I} \circ \boldsymbol{e}_\mathrm{I} - \nu(\boldsymbol{e}_\mathrm{II} \circ \boldsymbol{e}_\mathrm{II} + \boldsymbol{e}_\mathrm{III} \circ \boldsymbol{e}_\mathrm{III})]$$

gemessen. Im Experiment wird der Eigenwert σ_I gesteuert, und die Eigenwerte G_I und (nicht immer) $G_\mathrm{II} = G_\mathrm{III} = -\nu G_\mathrm{I}$ werden beobachtet. Unter der Annahme $\nu =$ const. wird von vielen Autoren vermutet, daß sich in diesem Versuch auch das Materialgesetz für nichtlinearelastische Stoffe bestimmen läßt. In [3] wird ein Verfahren vorgestellt, mit dem sich solche Aufgaben bei Kenntnis von $G_\mathrm{I} = f(\sigma_\mathrm{I})$ routinemäßig erledigen lassen sollen. Nimmt man aber an, daß ein duales nichtlineares Stoffgesetz durch die Tensorfunktion (2.14b) gegeben ist, so zeigt Gleichung (2.19), daß dies nur für $L = N = 0$ gelten kann. Offenbar läßt sich der Koeffizient N auch bei Messung einer Querkontraktion gemäß Gleichung (2.19)

$$\nu = -a_0 = \frac{-G_\mathrm{II}}{G_\mathrm{I}} = \frac{-G_\mathrm{III}}{G_\mathrm{I}} = \frac{-(\Lambda + L\sigma_\mathrm{I})}{\Lambda + 2\Gamma + (L + 2M)\,\sigma_\mathrm{I}} \approx \frac{-(\Lambda + L\sigma_\mathrm{I})}{\Lambda + 2\Gamma} \tag{2.29}$$

nicht aus dem einachsigen Spannungszustand bestimmen. Mit anderen Worten: Stoffe, die sich nur im Wert für N unterscheiden, können im einachsigen Zug-Druckversuch nicht mehr auseinandergehalten werden. Zur Bestimmung aller Koeffizienten in (2.14b) muß also zusätzlich zum einachsigen Spannungszustand mindestens ein weiterer Versuch durchgeführt werden. Dazu könnten z. B. die durch (2.18) oder (2.21) gegebenen Spezialfälle dienen. Welche Experimente man auswählt, wird von den gegebenen Möglichkeiten, den Kosten und der angestrebten Sicherheit in der Aussage abhängen.

Geht man von dem nichtlinearen Stoffgesetz gemäß der Approximation (2.14a) aus, so verläuft die Argumentation nicht ganz so einsichtig. Zunächst fällt auf, daß die für eine experimentelle Auswertung zuständigen Approximationsfunktionen wesentlich komplizierter gebaut sind und außer im hydrostatischen Fall (2.23) auf nichtlineare Approximationsaufgaben führen. Entwickelt man aber die Gleichung (2.24a) für den einachsigen Spannungszustand, so erhält man die Querkontraktion

$$\nu = -a_0 = \frac{\lambda + lG_\mathrm{I}}{\mu 2(\lambda + \mu) + (4l - 2m + n)\, G_\mathrm{I}} \approx \frac{\lambda}{2(\lambda + \mu)} + \frac{2l(\mu - \lambda) + \lambda(2m - n)}{4(\lambda + \mu)^2}\, G_\mathrm{I}. \tag{2.30}$$

Dies, in Gleichung (2.24b) eingesetzt, liefert unter Vernachlässigung von Termen höher als G_I^2

$$\sigma_\mathrm{I} = \left[\frac{\mu(3\lambda + 2\mu)}{\lambda + \mu} + \frac{4l(\lambda^2 - \lambda\mu + \mu^2) + 2m(\lambda^2 + 8\lambda\mu + 4\mu^2) + 3n\lambda^2}{4(\lambda + \mu)^2}\, G_\mathrm{I}\right] G_\mathrm{I}. \tag{2.31}$$

Damit hat man auch hier ein lineares Approximationsproblem und sieht leicht, daß zur vollständigen Bestimmung der Kennwerte λ, μ, l, m und n mindestens eines der oben genannten Experimente zusätzlich zum einachsigen Zugversuch durchzuführen ist.

Entsprechende Entwicklungen lassen sich auch für die weiteren oben genannten Experimente angeben, etwa für den gleichen zweiachsigen Zug- oder Druckversuch (2.26). Die Konvergenz der durchgeführten Entwicklungen läßt sich aber erst nach der Bestimmung der Stoffkoeffizienten prüfen. Man findet, daß die Approximation degressiven Materialverhaltens (geringere als lineare Spannungszuwächse; $W < W_c$) durch (2.14a) z. B. im Druckbereich auf nach oben geöffnete quadratische Parabeln führt.

Mit $p = \frac{1}{3} J_1$ gilt nämlich für den hydrostatischen Zustand nach (2.23)

$$p = [3\lambda + 2\mu + (9l + n)\, G_I]\, G_I\,. \tag{2.32}$$

Das ist eine Parabel mit den Scheitelpunktkoordinaten G_{Is}, p_s:

$$G_{Is} = -\frac{3\lambda + 2\mu}{2(9l + n)}\,, \qquad p_s = \frac{(3\lambda + 2\mu)^2}{2(9l + n)}\,.$$

Die Umkehrfunktion der skalaren Gleichung (2.32) ergibt sich als Lösung einer quadratischen Gleichung zu

$$G_I = \frac{3\lambda + 2\mu}{2(9l + n)}\left[-1 + S\sqrt{1 - \frac{4(9l + n)}{(3\lambda + 2\mu)^2}\,p}\right]$$

mit

$$S := \operatorname{sign}\left(\frac{3\lambda + 2\mu}{9l + n}\right).$$

Für $|p| \leqq \frac{1}{2}\,|p_s|$ kann man die Wurzel entwickeln und erhält

$$G_I = \frac{1}{3\lambda + 2\mu}\,p + \frac{9l + n}{(3\lambda + 2\mu)^3}\,p^2 + \dots\,.$$

Das ist aber wegen (2.7) gerade die skalare Gleichung (2.18) für den hydrostatischen Zustand und damit ein Merkmal für die Konvergenz der zu dem dualen Gesetz führenden Entwicklung.

2.4 Zur nichtlinearen Erweiterung der klassischen Moduln

Den speziellen Charakter der mit nichtlinear erweiterten klassischen Moduln formulierten Stoffgesetze zeigt das in [9] für kleine Verzerrungen vorgeschlagene nichtlinearelastische Gesetz. Dort wird, ausgehend von den Entkopplungsannahmen, daß der Kugeltensor von $\boldsymbol{\sigma}$ eine umkehrbar eindeutige Funktion allein des Kugeltensors von $\boldsymbol{G}$ ist und daß entsprechendes auch für die Deviatoren gilt, das Hookesche Gesetz nichtlinear erweitert. Für diesen Fall enthalten die aus dem einachsigen Spannungszustand gewonnenen Materialfunktionen z. B. E und ν die vollständige Information, und die dualen Materialgesetze lassen sich analytisch angeben.

Die oben bereitgestellten Gleichungen zeigen aber, daß man ohne derartig einschränkende Annahmen auf Widersprüche stößt, wenn man die Moduln der klassischen linearen Elastizitätstheorie nichtlinear erweitern will. Bildet man etwa mit (2.19a) einen nichtlinearen „Elastizitätsmodul"

$$E(\sigma_I) := \left(\frac{\mathrm{d}G_I}{\mathrm{d}\sigma_I}\right)^{-1} = \frac{1}{\Lambda + 2\Gamma + (2L + 4M)\,\sigma_I}$$

und mit (2.18) einen „Kompressionsmodul"

$$\varkappa(\sigma_I) := \frac{1}{3}\left(\frac{\mathrm{d}G_I}{\mathrm{d}\sigma_I}\right)^{-1} = \frac{1}{3[3\Lambda + 2\Gamma + (18L + 2N)\,\sigma_I]}\,,$$

so zeigt Gleichung (2.29), daß die Gleichungen der klassischen Theorie wie z. B.

$$\varkappa = \frac{E}{3(1 - 2\nu)}$$

nicht mehr gelten. Daher lassen sich auch alle auf solchen Beziehungen fußenden klassischen Ergebnisse für eine nichtlineare Theorie nicht nutzen.

Für die analoge Diskussion an Hand der Approximation (2.14a) möge ein Zitat aus [5] stehen: „It is impossible to predict the magnitude of the second-order effects in terms of the infinitesimal elasticities λ and μ. Some distinguished savants have deluded themselves into a contrary opinion." Derart strenge Aussagen sind zwar prinzipiell richtig, doch darf nicht übersehen werden, daß auch die Approximationen durch die isotropen Tensorfunktionen (2.14) für große $\boldsymbol{\sigma}$ bzw. $\boldsymbol{G}$ zu großen Fehlern und zu unsinnigen Ergebnissen führen können. Anscheinend sind Polynome nicht immer ein geeignetes Funktionensystem zur Approximation der Potentiale W und W_c.

Die Approximation einachsiger Experimente durch einachsige Stoffgesetze ist sicher empfehlenswert, wenn sie auf einachsige Anwendungen (nichtlineare Fachwerke, Seilnetze) beschränkt wird. Eine skalare Funktion $E(\sigma_I)$ läßt sich dafür leicht geeignet finden. Oft werden auch transzendente und gebrochen rationale Funktionen mit unterschiedlich vielen freien Parametern vorgeschlagen [1, 2], so daß einachsige Zug-Druckversuche bis in instabile Bereiche hinein hervorragend approximiert werden können.

Hier wurde aber gezeigt, daß dieser Genauigkeitsgewinn beim einachsigen Spannungszustand mit einem Verlust der Kenntnisse über mehrachsige Spannungszustände einhergeht. Daher scheint es sinnvoll, über neue Ansätze für τ_0, τ_1, τ_2 bzw. $\varepsilon_0, \varepsilon_1, \varepsilon_2$ in (2.5) nachzudenken. Speziell für elastische Stoffe bietet sich dazu als Hilfsmittel das Materialgesetz (2.9) in Form einer skalaren Potentialfunktion an.

3. Isotropes inkompressibles Material

3.1 Approximation gummiartiger Stoffe

Gummiartige Materialien, die bis zu extremen reversiblen Dehnungen von einigen hundert Prozent Volumenänderungen nur im Promillebereich zeigen, können oft als praktisch inkompressibel angesehen werden. Dadurch vereinfacht sich das Stoffgesetz. Als Verzerrungsmaß ist hier der rechte Cauchy-Green-Tensor $\boldsymbol{C}$ geeignet. Er hat die Spektralform

$$\boldsymbol{C} = C_{\mathrm{I}}\boldsymbol{e}_{\mathrm{I}} \circ \boldsymbol{e}_{\mathrm{I}} + C_{\mathrm{II}}\boldsymbol{e}_{\mathrm{II}} \circ \boldsymbol{e}_{\mathrm{II}} + C_{\mathrm{III}}\boldsymbol{e}_{\mathrm{III}} \circ \boldsymbol{e}_{\mathrm{III}} . \tag{3.1}$$

Inkompressibilität bedeutet für die kubische Hauptinvariante III

$$\mathrm{III} := \det \boldsymbol{C} = (\det \boldsymbol{F})^2 = 1 . \tag{3.2}$$

Dann sind im Experiment allein die lineare und die quadratische Hauptinvariante I und II steuerbar. Sie lauten wegen (3.2)

$$\mathrm{I} := \mathrm{Sp}\, \boldsymbol{C} = C_{\mathrm{I}} + C_{\mathrm{II}} + \frac{1}{C_{\mathrm{I}}C_{\mathrm{II}}}, \qquad \mathrm{II} := \frac{1}{2}(\mathrm{Sp}\, \boldsymbol{C}^2 - \mathrm{I}^2) = -C_{\mathrm{I}}C_{\mathrm{II}} - \frac{1}{C_{\mathrm{I}}} - \frac{1}{C_{\mathrm{II}}}. \tag{3.3 a, b}$$

Die Diskussion sei wieder auf hyperelastische Materialien beschränkt. Das spezifische elastische Potential $W = V(\mathrm{I}, \mathrm{II})$ hängt nicht von III ab. Bei seiner Variation ist zu berücksichtigen, daß nicht alle Komponenten von $\boldsymbol{C}$ unabhängig variierbar sind. Erweitert man V bzw. W mit dem Lagrange-Multiplikator $\frac{p}{2\varrho}$ zu

$$\tilde{V}(\mathrm{I}, \mathrm{II}, \mathrm{III}) = V(\mathrm{I}, \mathrm{II}) + \frac{p}{2\varrho}(\mathrm{III} - 1) , \qquad \tilde{W} = V(\mathrm{I}, \mathrm{II}, \mathrm{III}) , \tag{3.4}$$

so findet man durch Differenzieren das Stoffgesetz

$$\boldsymbol{\sigma} = 2\varrho \frac{\partial \tilde{W}}{\partial \boldsymbol{C}}, \qquad \boldsymbol{\sigma} = 2\varrho \left(\frac{\partial V}{\partial \mathrm{I}} - \mathrm{I}\frac{\partial V}{\partial \mathrm{I}}\right)\boldsymbol{1} + 2\varrho\left(\frac{\partial V}{\partial \mathrm{II}}\right)\boldsymbol{C} + p\boldsymbol{C}^{-1} . \tag{3.5 a, b}$$

Es hat die Form (2.1b):

$$\boldsymbol{\sigma} = \gamma_{0(\mathrm{I}, \mathrm{II})}\, \boldsymbol{1} + \gamma_{1(\mathrm{I}, \mathrm{II})}\, \boldsymbol{C} + p\boldsymbol{C}^{-1} . \tag{3.5c}$$

Der Lagrange-Multiplikator muß aus einem Randwertproblem berechnet werden; p ist als hydrostatischer Druck interpretierbar. Der besondere Charakter der Stoffgesetze für elastische, inkompressible Materialien wird verschleiert, sobald p berechnet ist.

Für den ebenen und den einachsigen Spannungszustand folgt aus $\sigma_{\mathrm{III}} = 0$

$$p = -C_{\mathrm{III}}\gamma_{0(\mathrm{I}, \mathrm{II})} - C_{\mathrm{III}}^2\gamma_{1(\mathrm{I}, \mathrm{II})} . \tag{3.6a}$$

Dabei ergibt sich C_{III} alleine aus der Inkompressibilitätsbedingung (3.2); für den ebenen Spannungszustand erhält man

$$C_{\mathrm{III}} = \frac{1}{C_{\mathrm{I}}C_{\mathrm{II}}} \tag{3.7a}$$

und für den einachsigen Spannungszustand (einfache Dehnung)

$$C_{\mathrm{III}} = C_{\mathrm{II}} = \frac{1}{\sqrt{C_{\mathrm{I}}}}. \tag{3.7b}$$

In beiden Fällen gewinnt man mit $\gamma_{-1(\mathrm{I}, \mathrm{II})} := p$ die Darstellung

$$\boldsymbol{\sigma} = \gamma_{0(\mathrm{I}, \mathrm{II})}\, \boldsymbol{1} + \gamma_{1(\mathrm{I}, \mathrm{II})}\, \boldsymbol{C} + \gamma_{-1(\mathrm{I}, \mathrm{II})}\, \boldsymbol{C}^{-1} . \tag{3.6b}$$

Sie läßt nicht erkennen, daß (3.6b) kein Stoffgesetz ist, weil die Gleichung (3.6a) aus den Randbedingungen gewonnen wurde. Wo dies nicht möglich ist, etwa bei dem hydrostatischen Spannungszustand, bleibt $\boldsymbol{\sigma}$ um den Term $p\boldsymbol{C}^{-1}$ unbestimmt. Gleichung (3.6b) ist nur für Argumente $\boldsymbol{C}$ definiert, die einer Inkompressibilitätsbedingung (3.7) genügen. Geht man auf Tensoren $\boldsymbol{\sigma}_e$ und $\boldsymbol{C}_e$ mit den Hauptinvarianten I_e, II_e in zwei Raumdimensionen über, so ergibt sich die reduzierte Form

$$\boldsymbol{\sigma}_e = \gamma_{e0}(\mathrm{I}_e, \mathrm{II}_e)\, \boldsymbol{1} + \gamma_{e1}(\mathrm{I}_e, \mathrm{II}_e)\, \boldsymbol{C}_e . \tag{3.6c}$$

Die mit inneren Bindungen (Zwangsbedingungen) eines Materials verbundenen Besonderheiten werden von dem in [3] vorgeschlagenen Verfahren nicht berücksichtigt.

3.2 Der einachsige Spannungszustand gummiartiger Stoffe

Nach dem oben Gesagten ist klar, daß die in einem speziellen Experiment wie dem einachsigen Zugversuch gewonnenen Spannungs-Dehnungs-Kurven nicht durch beliebige Ansatzfunktionen approximiert werden dürfen. Vielmehr wird der Funktionsverlauf durch die Randbedingungen und die Inkompressibilität wesentlich festgelegt. Auch lassen sich die etwa für die klassischen Moduln ermittelten Funktionen nicht auf andere Spannungs- oder Dehnungszustände übertragen, weil jeweils p erst aus der Lösung eines Randwertproblems folgt und eine spezielle Randbedingung bereits in den Moduln versteckt ist.

Zur Diskussion der klassischen Moduln geht man auf G_{I} über. Die von $C_{\mathrm{I}} = 1 + 2G_{\mathrm{I}}$ ausgehenden Reihenentwicklungen nach Potenzen von G_{I} konvergieren hier für $|G_{\mathrm{I}}| < \frac{1}{2}$ d. h. für $C_{\mathrm{I}} < 2$ (50% Dehnung).

Aus (3.7) ergibt sich als Poissonsche „Querkontraktionszahl" ν (Bild 1)

$$\nu = -\frac{C_{\mathrm{I}}^{-1/2} - 1}{C_{\mathrm{I}} - 1} = -\frac{(1 + 2G_{\mathrm{I}})^{-1/2} - 1}{2G_{\mathrm{I}}}, \qquad \nu_{\mathrm{appr}} = \tfrac{1}{2} - \tfrac{3}{4} G_{\mathrm{I}} + \tfrac{5}{4} G_{\mathrm{I}}^2 - \tfrac{35}{16} G_{\mathrm{I}}^3 + \tfrac{63}{16} G_{\mathrm{I}}^4 - \dots \tag{3.8 a, b}$$

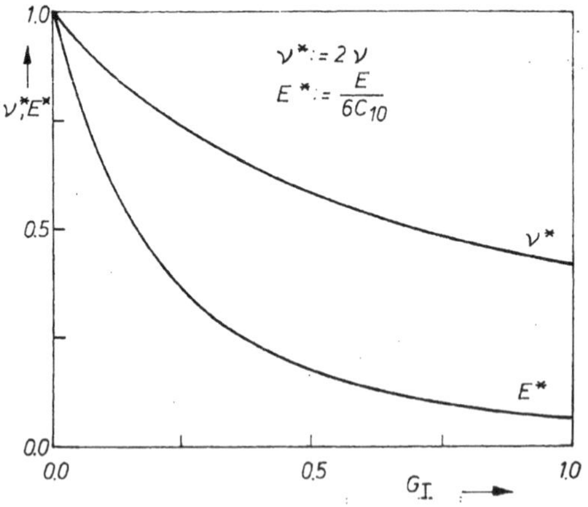

Bild 1. Einachsiger Zugversuch für das Neo-Hooke-Material: „Querkontraktionszahl" ν nach Gleichung (3.8a) und „Elastizitätsmodul" E nach Gleichung (3.11a)

ν ist für inkompressible Materialien eine stoffunabhängige rein kinematische Größe. Nur für verschwindende Dehnungen wird $\nu = \frac{1}{2}$. Für diesen Wert sind die klassischen Moduln λ und $\varkappa$ nicht definiert.

Der Youngsche Elastizitätsmodul E werde noch für zwei Beispiele diskutiert. Für das lineare Neo-Hooke-Material mit nur einem Stoffkoeffizienten C_{10} und dem Materialgesetz

$$\varrho V = C_{10}(\mathrm{I} - 3), \qquad \boldsymbol{\sigma} = 2C_{10}\boldsymbol{1} + p\boldsymbol{C}^{-1} \tag{3.9 a, b}$$

wird der einachsige Spannungszustand durch

$$\boldsymbol{\sigma} = 2C_{10}(1 - C_{\mathrm{I}}^{-3/2})\, \boldsymbol{e}_{\mathrm{I}} \circ \boldsymbol{e}_{\mathrm{I}} \tag{3.10}$$

beschrieben. Dann lautet der „Elastizitätsmodul" (Bild 1)

$$E = 6C_{10}C_{\mathrm{I}}^{-5/2} = 6C_{10}(1 + 2G_{\mathrm{I}})^{-5/2}, \qquad E_{\mathrm{appr}} = 6C_{10}(1 - 5G_{\mathrm{I}} + \tfrac{35}{2} G_{\mathrm{I}}^2 - \tfrac{105}{2} G_{\mathrm{I}}^3 + \tfrac{1155}{8} G_{\mathrm{I}}^4 - \dots). \tag{3.11 a, b}$$

Die Nichtlinearität ist hier rein kinematisch bedingt. Eine andere Approximationsfunktion zu suchen oder mehr Koeffizienten als C_{10} zu bestimmen, wäre wenig sinnvoll. Unterdrückt man die Nichtlinearität von ν nach (3.8b), so ist es nur konsequent, mit E nach (3.11b) ebenso zu verfahren.

Daran ändert auch der Übergang auf ein physikalisch nichtlineares Material nichts, dessen bekanntester Vertreter das Mooney-Rivlin-Material mit den zwei Koeffizienten C_{10}, C_{01} eines unvollständigen Polynoms ist. Dafür gilt

$$\varrho V = C_{10}(\mathrm{I} - 3) - C_{01}(\mathrm{II} + 3), \qquad \boldsymbol{\sigma} = 2(C_{10} + C_{01}\mathrm{I})\,\boldsymbol{1} - 2C_{01}\boldsymbol{C} + p\boldsymbol{C}^{-1}. \tag{3.12 a, b}$$

Für den einachsigen Spannungszustand ergibt sich

$$\boldsymbol{\sigma} = (2C_{10} + 2C_{01}C_{\mathrm{I}}^{-1/2})(1 - C_{\mathrm{I}}^{-3/2})\, \boldsymbol{e}_1 \circ \boldsymbol{e}_{\mathrm{I}}, \tag{3.13}$$

$$E = 6C_{10}C_{\mathrm{I}}^{-5/2} + 2C_{01}(4C_{\mathrm{I}}^{-3} - C_{\mathrm{I}}^{-3/2}), \tag{3.14 a}$$

$$E_{\mathrm{appr}} = 6(C_{10} + C_{01}) - 6C_{10}\,(5G_{\mathrm{I}} - \tfrac{35}{2} G_{\mathrm{I}}^2 + \tfrac{105}{2} G_{\mathrm{I}}^3 - \dots) - 6C_{01}\,(7G_{\mathrm{I}} - \tfrac{59}{2} G_{\mathrm{I}}^2 + \tfrac{605}{6} G_{\mathrm{I}}^3 - \dots). \tag{3.14 b}$$

Auch hier wird der Kurvenverlauf von dem Faktor $1 - C_{\mathrm{I}}^{-3/2}$ bestimmt. Beide Kurven (3.10) und (3.13) nähern sich für große C_{I} asymptotisch der maximalen Spannung $\sigma_{\mathrm{I}} = 2C_{10}$. Der Unterschied zwischen beiden Approximationen findet sich überraschenderweise gerade bei kleinen Dehnungen. Für reale Stoffe ist C_{01} meist wesentlich kleiner als C_{10}. Dann läßt sich C_{01} nicht hinreichend genau aus dem einachsigen Versuch ermitteln [10, 11].

4. Zusammenfassung

Eindimensionale Approximationen mechanischen Stoffverhaltens sind aus einachsigen Experimten zu gewinnen. Sie können an diese leicht angepaßt werden und beschreiben oft auch benachbarte mehrdimensionale Zustände hinreichend gut. Sie lassen sich aber grundsätzlich nicht zur Modellierung beliebiger mehrdimensionaler Zustandsänderungen nichtlinearer Stoffe nutzen. Um es kurz zu sagen: Bei Reduktion auf Probleme niedrigerer Dimension sind Ein- und Ausgabetensoren $\boldsymbol{X}$ und $\boldsymbol{Y}$ in (2.1a) Tensoren niedrigerer Dimension in dem Sinne, daß sie im ebenem Fall nur zwei und im einachsigen Fall nur eine unabhängige Invariante besitzen. Dann kann die Darstellung (2.1a) mit dem Satz von Cayley-Hamilton im ebenen Fall auf

$$\boldsymbol{Y}(\boldsymbol{X}) = \varphi_{20}\boldsymbol{1} + \varphi_{21}\boldsymbol{X} \tag{4.1a}$$

und im einachsigen Fall auf (die eigentlich skalare Gleichung)

$$\boldsymbol{Y}(\boldsymbol{X}) = \varphi_{10}\boldsymbol{1} \tag{4.1b}$$

reduziert werden. Was von „Tensoriellen Verallgemeinerungen" von (4.1b) versprochen, aber für wirkliche Materialien höchstens ausnahmsweise gehalten wird, ist, daß $\varphi_0 = \varphi_0(\varphi_{10})$, $\varphi_1 = \varphi_1(\varphi_{10})$ und $\varphi_2 = \varphi_2(\varphi_{10})$ sein soll. An Hand von Beispielen für Stoffgesetze wird gezeigt, daß selbst die Verallgemeinerung von (4.1a) zu (2.1a) nur bei sehr speziell aufgebauten Materialgesetzen gelingt. Offenbar ist die Reduktion der dreidimensionalen Stoffgleichung (2.1) auf Gleichungen geringerer Dimension (4.1) i. a. nicht in eindeutiger Weise zurücknehmbar.

Für kompressible, elastische Stoffe wird gezeigt, daß einachsige Approximationen grundsätzlich nicht mehr die vollständige Information des dreiachsigen Stoffgesetzes enthalten. Indessen enthält die einachsige Approximation inkompressibel elastischer Stoffe zusätzliche Information aus den Randbedingungen.

Parallel werden duale Materialgesetze für kompressible Stoffe diskutiert. Diese existieren für elastische Stoffe zwar per Definition, doch lassen sie sich nur im linearen Fall analytisch angeben. Es werden Wege aufgezeigt, die trotzdem zu dualen Formulierungen approximativer Gesetze führen. Nichtlineare duale Materialgesetze werden u. a. in der nichtlinearen Photoelastizität [12] und zum Aufstellen komplementärer (konjugierter) Variationsprinzipe [13] benötigt.

Literatur

1 CHEN, W. F.; SALEEB, A. F.: Constitutive equations for engineering materials. Vol. I: Elasticity and modelling. Wiley, New York 1981.
2 EIBL, J. et al.: Concrete under multiaxial states of stress-constitutive equations for practical design. Bull. d'Information, Nr. 156, Comité Euro-international du Béton (CEB), Juin 1983.
3 BETTEN, J.: Beitrag zur tensoriellen Verallgemeinerung einachsiger Stoffgesetze. ZAMM 66 (1986), 577–581.
4 TRUESDELL, C.: The mechanical foundations of elasticity and fluid dynamics. J. Rat. Mech. Anal. 1 (1952), 125–300. Corrected reprint, Intl. Sci. Rev. Ser., Gordon & Breach, New York 1965.
5 TRUESDELL, C.; NOLL, W.: Die nicht-linearen Feldtheorien der Mechanik. Handbuch der Physik, Bd. III/3. Springer, Berlin 1965.
6 PFLEIDERER, H.; SEEGER, A.; KRÖNER, E.: Nichtlineare Elastizitätstheorie geradliniger Versetzungen. Z. Naturforschung 15a (1960), 758–772.
7 SEEGER, A.; BUCK, O.: Die experimentelle Ermittlung der elastischen Konstanten höherer Ordnung. Z. Naturforschung 15a (1960), 1056–1067.
8 WANG, C. C.; TRUESDELL, C.: Introduction to rational elasticity. Noordhoff, Leyden 1973.
9 KAUDERER, H.: Über ein nichtlineares Elastizitätsgesetz. Ing.-Arch. 17 (1949), 450–480.
10 HAINES, D. W.; WILSON, W. D.: Strain-energy density function for rubber-like materials. J. Mech. Phys. Solids 27 (1979), 331–343.
11 MAROTZKE, CH.: Bewertung von Stoffgesetzen für Elastomere anhand von Versuchen und numerischen Untersuchungen. ZAMM 65 (1985), 388–390.
12 LAERMANN, K. H.: Über eine Theorie der nichtlinearen Photoelastizität. ZAMM 66 (1986), T 161–T 163.
13 KOITER, W. T.: On the principle of stationary complementary energy in the nonlinear theory of elasticity. Siam. J. Appl. Math. 25 (1973), 424–434.

Eingegangen am 9. November 1987

Anschrift: Dipl.-Ing. M. STAAT, Prof. Dr.-Ing. J. BALLMANN, Lehr- und Forschungsgebiet Mechanik der RWTH Aachen, Templergraben 64, D-5100 Aachen, BRD

BOOK REVIEWS

Wallisch, W.; Hermann, M., Numerische Behandlung von Fortsetzungs- und Bifurkationsproblemen bei Randwertaufgaben. Leipzig, B. G. Teubner Verlagsgesellschaft 1987. 188 S., M 19,50. ISBN 3-322-00431-7 (Teubner-Texte zur Mathematik 102)

Das Buch behandelt die numerische Lösung von parameterabhängigen Randwertaufgaben bei gewöhnlichen Differentialgleichungen der Form

$$\left.\begin{aligned} y' &= f(t, y, \lambda), \quad a \leqq t \leqq b \\ r(y(a), y(b), \lambda) &= 0 \end{aligned}\right\} \tag{1}$$

mit Hilfe von Fortsetzungsmethoden. Dabei wird auf die bekannte Klasse von Schießverfahren (u. a. Mehrziel-Verfahren) Bezug genommen. Die Voraussetzung solcher Methoden ist, daß die zu bestimmende Lösung „isoliert" ist, d. h. die Frechet-Ableitung des zu (1) gehörenden Operators in der gesuchten Lösung invertierbar ist. Dies ist bei den allermeisten Lösungen der Fall, jedoch gerade nicht in einigen besonders interessanten Fällen, den sogenannten Verzweigungspunkten.

Die Kapitel I und II beschäftigen sich mit der grundlegenden Theorie der Verzweigung in endlich-dimensionalen Räumen sowie bei Randwertproblemen gewöhnlicher Differentialgleichungen. Diese Probleme sind ja nah verwandt, da die verschiedenen Schießverfahren die Zurückführung des unendlich-dimensionalen Problems in ein endlich-dimensionales erlauben.

Kap. III behandelt funktionalanalytische Methoden im Zusammenhang mit dem Ursprungsproblem (1) und Varianten davon. Das Prinzip aller dort angegebenen Verfahren ist die Einbettung des Problems in ein höher-dimensionales System unter Heranziehung zusätzlicher Informationen wie Nullvektoren der Linearisierung. Dabei ergibt sich fast immer ein reguläres Problem, parametrisiert durch einen Störungsparameter s, so daß man die Lösungen $y(s)$, $\lambda(s)$ von (1) in dieser Form numerisch in gutartiger Form mittels Schießverfahren bestimmen kann. Die Diskussion umfaßt Punkte wie Hopf-Verzweigung, gestörte Verzweigung, Bestimmung von Verzweigungspunkten und Umkehrpunkten und Falten.

Das Buch ist gut lesbar geschrieben und bietet eine Fülle von Informationen, die man in dieser Zusammenstellung sonst nur in der Originalliteratur findet. Ein FORTRAN-Programm zur Kurvenverfolgung ist mit abgedruckt.

Wiesbaden H. Weber

Engeln-Müllges, G.; Reutter, F., Formelsammlung zur numerischen Mathematik mit C-Programmen. Mannheim etc., B.I.-Wissenschaftsverlag 1987, XV, 564 S., DM 48,–. ISBN 3-411-03112-3

Mit ihrem Buch wenden sich die Autoren an Ingenieure, Informatiker und Naturwissenschaftler, die Standardverfahren der Numerischen Mathematik benötigen. Eine entsprechende Ausgabe des Buches mit FORTRAN 77 Programmen existierte bereits. Die Aufgabenstellungen wesentlicher Teilgebiete der Numerischen Mathematik und die Methoden zu ihrer Lösung sind in Form von Definitionen, Sätzen, Beschreibungen und Algorithmen in übersichtlicher Weise dargestellt. Auf Beweise wird, dem Charakter des Buches entsprechend, verzichtet. Der an einem tieferen Verständnis interessierte Leser wird durch detaillierte Literaturhinweise unterstützt.

Für viele der im Textteil beschriebenen Methoden befinden sich C-Programme im Anhang. Sie sind nicht voll in die Darstellung der Textteile integriert. Das bezieht sich sowohl auf die Bezeichnungen in den Algorithmen als auch auf die Ausführlichkeit in der Behandlung komplizierterer Lösungsverfahren. Die C-Programme sind umfassend dokumentiert. Testbeispiele mit Ergebnissen stehen i. allg. nicht bereit.

Die einzelnen Kapitel sind so aufgebaut, daß ausgehend von einfachen Verfahren zu komplizierteren übergegangen wird. Obwohl vergleichende Aussagen zu verschiedenen Methoden für gleiche Aufgabenstellungen nicht fehlen, wird auf bestimmte Schwächen herkömmlicher Verfahren nicht immer in aller Konsequenz hingewiesen. Entsprechend ihrer grundlegenden Bedeutung wird den Verfahren zur Lösung linearer Gleichungssysteme ein breiter Raum gewidmet. Die modernen Methoden zur Konditionszahlschätzung haben keine Aufnahme in das Buch gefunden. Im Abschnitt Eigenwerte und -vektoren wird das spezielle Eigenwertproblem behandelt. Gut bekannte, weiterführende, auch in Buchform vorliegende Programmpakete zur Linearen Algebra (LINPACK, EISPACK), die wesentliche zusätzliche Algorithmen (lineare Ausgleichsrechnung, Singulärwertzerlegung, Eigenwertaufgaben) enthalten, werden zitiert, aber nicht hervorgehoben. In den Abschnitten Nichtlineare Gleichungen und Gleichungssysteme, Approximation, Interpolation und Differentiation sind die wesentlichen Standardverfahren dargestellt. Besonderes Augenmerk wird der Interpolation mit Hilfe verschiedener ein- und zweidimensionaler Spline-Funktionen gewidmet. Im Kapitel Quadratur ist eine große Zahl von Quadraturformeln zusammengestellt worden. Programme sind für diesen Abschnitt nicht vorhanden. Hier wäre ein gezielter Hinweis auf bekannte Programmpublikationen (etwa QUADPACK), die auch die Verfahrensauswahl unterstützen, angebracht. Bezüglich der Verfahren für Anfangswertaufgaben expliziter gewöhnlicher Differentialgleichungen existieren getrennte Kapitel für skalare und Systeme von Differentialgleichungen. Das Problem steifer Differentialgleichungen hätte auch in einer Formelsammlung deutlicher herausgearbeitet werden müssen. Für die Lösung von Zweipunkt-Randwertaufgaben sind das einfache Schießverfahren (mit Programm), die Mehrzielmethode und Differenzenverfahren in das Buch aufgenommen worden. Ein Kapitel Extremalaufgaben könnte noch zur Abrundung des Buches beitragen.

Berlin G. Hebermehl

Bongaarts, P. J. M.; Kerf, E. A. de; Kersten, P. H. M., Proceedings Seminar 1984–1986. Mathematical Structures in Field Theories. Vol. 1. Amsterdam, Stichting Mathematisch Centrum 1988, III, 137 pp., Dfl. 21.60. ISBN 90-6196-345-1 (CWI Syllabus 16)

Das vorliegende Buch ist der erste Teil eines Berichtes über ein Seminar zur mathematischen Struktur der Quantenfeldtheorie, das zwischen 1984 und 1986 an der Universität Amsterdam abgehalten wurde. Die drei in ihm enthaltenen Beiträge stehen nur in einem losen Zusammenhang und behandeln durchaus verschiedene Aspekte der Feldtheorie.

Der Teil „The mathematical structure of free quantum fields. Gaussian systems" von Bongaarts nimmt die Wightmansche Formulierung zum Ausgangspunkt. Die freien Felder werden durch die Eigenschaft ihrer n-Punktfunktionen, durch Trunkierung auf die 2-Punktfunktion zurückführbar zu sein, gekennzeichnet („Gaußsche Systeme"). Das Klein-Gordon-Feld wird von diesem Ansatz bis zur kanonischen Quantisierung geführt. Ein Anhang referiert die Fock-Quantisierung.

Der Beitrag „Particles, fields and scattering in relativistic quantum theory" von de Kerf behandelt vor allem den kinetischen Teil der Streutheorie. Ausgehend von der Analyse der Poincaregruppe und ihrer Darstellungen werden die Transformationseigenschaften von Feldoperatoren, Erzeugungs- und Vernichtungsoperatoren und Vielteilchenzuständen besprochen. Die folgende knappe Einführung in die Streutheorie orientiert sich am LSZ-Formalismus.

Der letzte Teil, „Infinitesimal symmetries of partial differential equations with an application to the construction of the instanton and monopole solution of the self-dual Yang-Mills equations", der von Kersten geschrieben wurde, erklärt sich hinreichend durch seine Überschrift. Seine Besonderheit besteht in einer sehr konzentrierten beweislosen Einführung in die Theorie der Jet-Bündel, der auf ihnen erklärten Cartanschen Formenideale und Lie-Bäcklund-Symmetrien sowie in den ausführlich vorgerechneten Beispielen.

Die Autoren haben versucht, auf wenigen Seiten Einführungen in umfangreiche theoretische Gebäude zu geben. Nimmt man die zahlreichen hierdurch bedingten Verkürzungen in Kauf, so wird das Buch dieser Aufgabe etwa gerecht.

Leipzig A. Uhlmann

ZAMM · Z. angew. Math. Mech. 69 (1989) 2, 83 – 92

JUNGHANNS, P.; OESTREICH, D.

Numerische Lösung des Staudammproblems mit Drainage

Es wird für das zweidimensionale Staudammproblem mit Drainage bei linearer oder gekrümmter Dammkurve ein numerisches Verfahren angegeben, um den freien Rand (Sickerlinie) zu bestimmen. Das Verfahren beruht auf der näherungsweisen Lösung äquivalenter singulärer Integralgleichungen auf dem Intervall [−1, 1] mittels Interpolationspolynome. Die numerischen Ergebnisse zeigen, daß die Durchflußrate und die Sickerlinie nur wenig vom Dammprofil abhängen.

A numerical approach to determine the free surface for a two-dimensional seepage flow problem through a linear or a nonlinear staped dam with a toe drain is given. The method is based on the approximation of equivalent singular integral equations on the interval [−1, 1] with interpolation polynoms. Numerical results show that the flowrate and the free surface depend only weakly on the shape of the dam.

Для задачи о линейной или нелинейной плотине с дренажем задается вычислительной метод для определения депрессионной кривой. Основа этого метода является приближенное решение эквивалентных сингулярных интегральных уравнений на отрезке [—I, I] с помощью интерполяционных многочленов. Вычислительные результаты показывают, что расход и депрессионная кривая только мало зависят от профиля плотины.

Für das zweidimensionale „Drainage-Problem" bei gekrümmter Dammkurve wurde in [21] durch Transformation mittels konformer Abbildung in eine im allgemeinen nichtlineare singuläre Integralgleichung die eindeutige Existenz der Lösung gezeigt. Die Gleichung des freien Randes (Sickerlinie) konnte in einer Parameterdarstellung angegeben werden. Die vorliegende Arbeit beschäftigt sich nun im linearen und nichtlinearen Fall mit der numerischen Bestimmung der Sickerlinie und der Durchflußrate, die für praktische Zwecke von besonderer Bedeutung sind. Dabei werden die in den letzten Jahren erzielten Ergebnisse bei der numerischen Behandlung linearer singulärer Integralgleichungen auf dem Intervall (siehe z. B. [4], [5], [11]—[18]) nutzbar gemacht.

Bisher wurde der freie Rand auf der Grundlage von Variationsungleichungen und Finite-Elemente- bzw. Finite-Differenzen-Methoden (siehe [24] und die dortige Bibliographie) berechnet. Mit dem hier vorgestellten Verfahren werden gleichwertige Resultate gewonnen, der rechentechnische Aufwand, insbesondere der Speicherplatzbedarf, dürfte jedoch im allgemeinen geringer sein. Die „klassische" Methode der konformen Abbildung (siehe z. B. [23]), bei der der freie Rand in analytischer Form beschrieben wird, ist dagegen nur für den linearen Fall und da auch nur bei gewissen Kleinheitsforderungen an das Verhältnis der Parameter anwendbar.

Das in der vorliegenden Arbeit beschriebene Verfahren läßt sich auch auf formal ähnliche Probleme, z. B. die Technologie der elektrochemischen Bearbeitung ([22]), übertragen.

1. Problemstellung

Wir betrachten dasselbe Modell wie in [21] (vgl. auch [1], [24], [26]). Das Strömungsgebiet G wird begrenzt durch einen Teil der Profilkurve der Stirnwand des Staudamms OA: $y = f(x)$, der Sickerlinie AB: $y = \varphi(x)$ (dem freien Rand) und der Dammbasis BO: $y = 0$, die sich unterteilt in den undurchlässigen Teil des Dammbodens LO und die Drainage BL (vgl. Abb. 1). Die Funktion $f(x)$ wird als hölderstetig differenzierbar angenommen, und ihre Ableitung genüge den Abschätzungen

$$0 < m_0 \leqq f'(x) \leqq M_0 < +\infty\,, \qquad x \in [0, h]\,, \tag{1}$$

mit gewissen Konstanten m_0 und M_0. Außerdem sei $f(0) = 0$ und $f(h) = H$, wobei H die Höhe des Wasserspiegels vor dem Damm bedeutet.

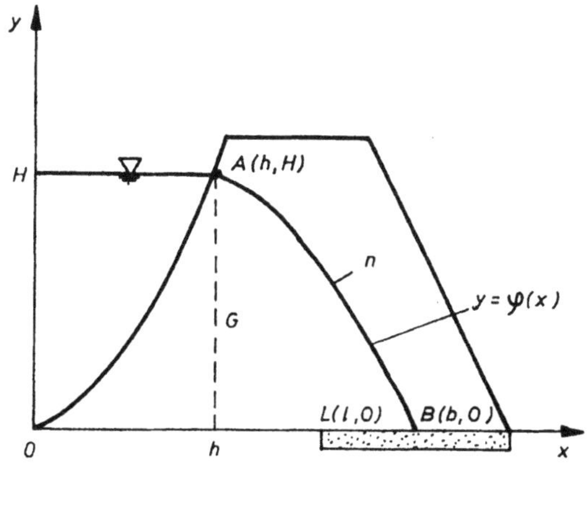

Abb. 1

Das zugehörige freie Randwertproblem ist der nichtlinearen singulären Integralgleichung

$$x(\tau) - \frac{1}{\pi} \int_1^{1/k} \frac{f(x(\sigma))\,d\sigma}{\sigma - \tau} - D = \frac{1}{\pi} \int_{-1}^{1} \frac{r(k, \sigma)\,d\sigma}{\sigma - \tau}, \tag{2}$$

$\tau \in (1, 1/k)$, mit den Randbedingungen

$$x(1) = h\,, \qquad x(1/k) = 0 \tag{3}$$

äquivalent. Dabei bezeichnen D eine reelle Zahl und $r(k, \tau)$ die Funktion

$$r(k, \tau) = \frac{H}{2} + \frac{H}{2K} F(k, \tau) \tag{4}$$

mit dem elliptischen Integral erster Gattung

$$F(k, \tau) = \int_0^{\tau} \frac{d\sigma}{\sqrt{(1 - k^2\sigma^2)(1 - \sigma^2)}}, \qquad 0 \leqq k \leqq 1\,, \tag{5}$$

und dem vollständigen elliptischen Integral erster Gattung

$$K = K(k) = F(k, 1)\,.$$

Für fixiertes $k \in (0, 1)$ hat das Problem (2), (3) eine eindeutige Lösung

$$(x, D) \in \bigcup_{p>1} W_p^1(1, 1/k) \times R\,.$$

Der Parameter k wird durch die Bedingung (vgl. [21])

$$\frac{1}{\pi} \int_{-1}^{1} \frac{r(k, \sigma)\,d\sigma}{\sigma + 1/k} + \frac{1}{\pi} \int_1^{1/k} \frac{f(x(\sigma))\,d\sigma}{\sigma + 1/k} + D = l \tag{6}$$

eindeutig festgelegt.

2. Algorithmus zur Bestimmung der Sickerlinie und der Durchflußrate

Sei $g = f^{-1}$ die Umkehrfunktion zu $y = f(x)$. Wir transformieren die Integralgleichung in eine Gleichung über dem festen Intervall $(-1, 1)$ mit homogenen Randbedingungen, wobei wir $y(\tau) = f(x(\tau))$ als die gesuchte Größe betrachten. Mittels

$$y(\tau) = u(t) + y_0(t)\,, \qquad y_0(t) = H(1 - t)/2\,, \qquad \tau = T(t) = [(1 - k)\,t + 1 + k]/2k\,, \qquad \sigma = T(s) \tag{7}$$

erhalten wir aus (2) und (3) die singuläre Integralgleichung

$$g[u(t) + y_0(t)] - \frac{1}{\pi} \int_{-1}^{1} \frac{u(s)\,ds}{s - t} - D = R(t)\,, \tag{8}$$

$t \in (-1, 1)$, mit den Randbedingungen

$$u(-1) = u(1) = 0\,. \tag{9}$$

Außerdem wird durch den Ansatz (7) die schwache Singularität in der rechten Seite von (2) (im Punkt $\tau = 1$) behoben. Die rechte Seite in (8) hat nach partieller Integration die Gestalt

$$R(t) = H[(1 - t)\ln(1 - t) + (1 + t)\ln(1 + t) + 2\ln[(1 - k)/2k] - 2]/2\pi -$$
$$- \frac{H}{2K\pi} \int_{-1}^{1} \frac{\ln[T(t) - \sigma]\,d\sigma}{\sqrt{1 - k^2\sigma^2}\sqrt{1 - \sigma^2}}. \tag{10}$$

Bemerkung 1: Wenn wir die Lösung von (8) in der Form $u(t) = \omega(t)\,v(t)$ mit einer hölderstetigen Funktion $\omega(t) > 0$, $t \in (-1, 1)$, und $\omega(-1) = \omega(1) = 0$ suchen, werden die Randbedingungen (9) offenbar erfüllt.

Die Bestimmungsgleichung (6) für den Parameter $k \in (0, 1)$ geht über in

$$L(k) := \frac{H}{\pi}\left[\ln\left(1 + \frac{1}{k}\right) + \frac{2}{1 - k}\ln\frac{2}{1 + k} - 1\right] - \frac{H}{2K\pi} \int_{-1}^{1} \frac{\ln(\sigma + 1/k)\,d\sigma}{\sqrt{1 - k^2\sigma^2}\sqrt{1 - \sigma^2}} +$$
$$+ \frac{1}{\pi} \int_{-1}^{1} \frac{u(s)\,ds}{s + \dfrac{3 + k}{1 - k}} + D - l = 0\,, \tag{11}$$

wobei (u, D) die eindeutige Lösung von (8), (9) bezeichnet. Somit läßt sich die Sickerlinie (der freie Rand AB) aus der Parameterdarstellung (vgl. [21], Satz 2)

$$x(t) = \frac{H}{\pi}\left[\frac{1-kt}{1-k}\ln\left(\frac{1-kt}{k}\right) - \frac{k(1-t)}{1-k}\ln(1-t) - 1\right] -$$

$$-\frac{H}{2K\pi}\int_{-1}^{1}\frac{\ln|\sigma - t|\,d\sigma}{\sqrt{1-k^2\sigma^2}\sqrt{1-\sigma^2}} + \frac{1}{\pi}\int_{-1}^{1}\frac{u(s)\,ds}{s - \frac{2kt}{1-k} + \frac{1+k}{1-k}} + D\,, \tag{12}$$

$$y(t) = r(k, t)\,, \qquad t \in (-1, 1)\,,$$

berechnen. Die Durchflußrate q erhält man aus

$$q = \frac{K(k')}{2K(k)}H\,, \qquad k' = \sqrt{1-k^2}\,. \tag{13}$$

Wir beschreiben nun den Algorithmus zur numerischen Lösung des Problems (8), (9), (11). Wegen $L(0) = -\infty$, $L(1) = +\infty$ und der strengen Monotonie von $L(k)$ (siehe [21]) bietet sich zur näherungsweisen Bestimmung von $k \in (0, 1)$ eine regula falsi an. Wir nehmen an, daß wir in Anlehnung an die Bemerkung 1 eine Gaußsche Quadraturformel

$$\frac{1}{\pi}\int_{-1}^{1}\omega(t)\,v(t)\,dt \sim \sum_{i=1}^{n}\lambda_i v(t_i)$$

und ein Verfahren zur Berechnung der $n + 1$ Größen $\xi_i \sim v(t_i)$, $i = 1, \ldots, n$, und D aus (8) bei gegebenem $k \in (0, 1)$ zur Verfügung haben (vgl. im linearen Fall Abschnitt 3, im nichtlinearen Fall Abschnitt 4). Dann gehen wir bei der Berechnung der Sickerlinie und der Durchflußrate wie folgt vor.

A. Gib k_i, $i = 1, 2$, mit $0 < k_1 < k_2 < 1$ vor.

B. Berechne $L(k)$ für $k = k_1, k_2$ nach (11). Approximiere dabei

$$\frac{1}{\pi}\int_{-1}^{1}\frac{\ln(\sigma + 1/k)\,d\sigma}{\sqrt{1-k^2\sigma^2}\sqrt{1-\sigma^2}} \sim \frac{1}{m}\sum_{i=1}^{m}\frac{\ln(\sigma_i + 1/k)}{\sqrt{1-k^2\sigma_i^2}} \tag{14}$$

mit den Nullstellen der Tschebyscheff-Polynome erster Art

$$\sigma_i = \cos\frac{2i-1}{2m}\pi\,, \qquad i = 1, \ldots, m\,, \tag{15}$$

und

$$\frac{1}{\pi}\int_{-1}^{1}\frac{u(s)\,ds}{s + \frac{3+k}{1-k}} \sim \sum_{i=1}^{n}\frac{\lambda_i\xi_i}{s_i + \frac{3+k}{1-k}}\,. \tag{16}$$

C. Falls $|L(k_i)| < \varepsilon$ (vorgegebene Abbruchschranke) für ein i, setze $k := k_i$ und gehe zu F.

D. Falls $L(k_1) < 0 < L(k_2)$, so gehe zu E, sonst setze $k_2 := k_1$, wenn $L(k_1) > 0$ ist, und wähle ein neues $k_1 < k_2$, bzw. setze $k_1 := k_2$, wenn $L(k_2) < 0$ ist, und wähle ein neues $k_2 > k_1$. Gehe zu B.

E. Berechne

$$k = [k_1L(k_2) - k_2L(k_1)]/[L(k_2) - L(k_1)] \tag{17}$$

und $(\xi_i, i = 1, \ldots, n, D)$ für dieses k sowie $L(k)$ (nach B). Falls $|L(k)| < \varepsilon$ gehe zu F. Sonst setze $k_1 := k$, falls $L(k) < 0$, bzw. $k_2 := k$, falls $L(k) > 0$. Gehe zu E.

F. Berechne die Durchflußrate q nach (13) und die Sickerlinie nach (12) für die Parameterwerte $t = \tau_j = \cos(j\pi/m)$, $j = 1, \ldots, m - 1$ (Nullstellen der Tschebyscheff-Polynome zweiter Art). Approximiere dabei

$$\frac{1}{\pi}\int_{-1}^{1}\frac{\ln|\sigma - \tau_j|\,d\sigma}{\sqrt{1-k^2\sigma^2}\sqrt{1-\sigma^2}} \sim \frac{1}{m}\sum_{i=1}^{m}\frac{\ln|\sigma_i - \tau_j|}{\sqrt{1-k^2\sigma_i^2}} \tag{18}$$

und

$$\frac{1}{\pi}\int_{-1}^{1}\frac{u(s)\,ds}{s - \frac{2k}{1-k}\tau_j + \frac{1+k}{1-k}} \sim \sum_{i=1}^{n}\frac{\lambda_i\xi_i}{s_i - \frac{2k}{1-k}\tau_j + \frac{1+k}{1-k}}\,. \tag{19}$$

Bemerkung 2: Die elliptischen Integrale (5) und (6) lassen sich gleichfalls mittels der Tschebyscheff-Quadratur berechnen. Im allgemeinen ist jedoch deren Bestimmung auf der Grundlage der Methode des arithmetisch-geometrischen Mittels [10] effektiver, die wir deshalb im Programm benutzten.

3. Lösung der linearen singulären Integralgleichung

Gegeben sei die Funktion $g(y) = \frac{h}{H} y$, $y \in [0, H]$. Dann geht (8) in eine lineare singuläre Integralgleichung mit konstanten Koeffizienten über,

$$\frac{h}{H} u(t) - \frac{1}{\pi} \int_{-1}^{1} \frac{u(s)\, ds}{s - t} - D = R_0(t), \qquad t \in (-1, 1). \tag{20}$$

Dabei ist

$$R_0(t) = R(t) - \frac{h}{2}(1 - t).$$

Wir bestimmen die Lösung von (19) in der Gestalt

$$u(t) = \omega(t)\, v(t), \qquad \omega(t) = \frac{(1 + t)^\alpha (1 - t)^\beta}{\sqrt{1 + (h/H)^2}}, \qquad \beta = \frac{1}{\pi} \arctan \frac{H}{h}, \qquad \alpha = 1 - \beta, \tag{21}$$

(vgl. Bemerkung 1) mittels des Gaußschen Quadraturformelverfahrens (siehe z. B. [5], [11]–[13], [16]–[18]). Wir verwenden die Gaußschen Quadraturformeln

$$\frac{1}{\pi} \int_{-1}^{1} \omega(t)\, p(t)\, dt \sim \sum_{i=1}^{n} \lambda_i p(s_i), \qquad \frac{1}{\pi} \int_{-1}^{1} \chi(t)\, p(t)\, dt \sim \sum_{j=1}^{n+1} \mu_j p(t_j), \tag{22a, b}$$

mit $\chi(t) = \frac{(1 + t)^{-\alpha} (1 - t)^{-\beta}}{\sqrt{1 + (h/H)^2}}$. Dabei sind $\{s_i\}_{i=1}^{n}$ bzw. $\{t_j\}_{j=1}^{n+1}$ die Nullstellen der Jacobi-Polynome $P_n^{(\alpha,\beta)}$ bzw. $P_{n+1}^{(-\alpha,-\beta)}$. Wir suchen die Näherungslösung für $v(t)$ in der Form

$$v_n(t) = \sum_{i=1}^{n} \xi_i \prod_{\substack{l=1 \\ l \neq i}}^{n} \frac{t - s_l}{s_i - s_l} \tag{23}$$

und erhalten durch Anwendung des Quadraturformelverfahrens für die gesuchten Werte ξ_i, $i = 1, \dots, n$, und D das lineare algebraische Gleichungssystem

$$\sum_{i=1}^{n} \frac{\lambda_i \xi_i}{t_j - s_i} - D = \eta_j := R_0(t_j), \qquad j = 1, \dots, n + 1. \tag{24}$$

Aus (24) folgt sofort (vgl. [4], [5], [14])

$$\xi_i = \sum_{j=1}^{n+1} \frac{\mu_j \eta_j}{t_j - s_i}, \qquad i = 1, \dots, n, \qquad D = -\sum_{j=1}^{n+1} \mu_j \eta_j \Big/ \sum_{j=1}^{n+1} \mu_j. \tag{25a, b}$$

Für das beschriebene Diskretisierungsverfahren gilt die Fehlerabschätzung

$$(\|v - v_n\|_\omega^2 + |D|^2)^{1/2} \leqq \|R_0^{(n)} - R_0\|_\chi$$

mit

$$\|v\|_\omega^2 = \frac{1}{\pi} \int_{-1}^{1} \omega(t)\, |v(t)|^2\, dt, \qquad R_0^{(n)}(t) = \sum_{j=1}^{n+1} \eta_j \prod_{\substack{i=1 \\ i \neq j}}^{n+1} \frac{t - t_i}{t_j - t_i}$$

(vgl. [12], [13], [15]–[18]). Die rechte Seite in (24) wird analog zu (14), (15) approximiert. Somit ist zur näherungsweisen Lösung von (20) nur noch die Berechnung der Stützstellen und Gewichte der Quadraturformeln (22) erforderlich.

Die Systeme $\{s_i\}_{i=1}^{n}$ bzw. $\{t_j\}_{j=1}^{n+1}$ sind die Eigenwerte der Matrizen (siehe z. B. [8])

$$\begin{bmatrix} d_1 & e_1 & & & \mathbf{0} \\ e_1 & d_1 & \ddots & & \\ & \ddots & \ddots & \ddots & \\ & & \ddots & d_{n-1} & e_{n-1} \\ \mathbf{0} & & & e_{n-1} & d_n \end{bmatrix} \quad \text{bzw.} \quad \begin{bmatrix} d_0 & e_0 & & & \mathbf{0} \\ e_0 & d_1 & \ddots & & \\ & \ddots & \ddots & \ddots & \\ & & \ddots & d_{n-1} & e_{n-1} \\ \mathbf{0} & & & e_{n-1} & d_n \end{bmatrix}$$

mit

$$d_j = \frac{\alpha - \beta}{4j^2 - 1}, \quad j = 0, 1, \dots, n; \qquad e_j = \frac{\sqrt{(j + \alpha)(j + \beta)}}{2j + 1}, \quad j = 1, \dots, n - 1; \qquad e_0 = \sqrt{2(1 - \alpha)(1 - \beta)}. \tag{26 a; b; c}$$

Die Größen (26) sind die Koeffizienten in der (symmetrischen) dreigliedrigen Rekursionsformel für die entsprechenden Jacobi-Polynome. Die Eigenwerte werden mittels des QL-Algorithmus ([2]) berechnet. Die Gewichte μ_j, $j = 1, \dots, n+1$, genügen dem Gleichungssystem (siehe z. B. [17], Formel (3))

$$\sum_{j=1}^{n+1} \frac{\mu_j \tilde{P}_1^{(-\alpha,-\beta)}(t_j)}{t_j - s_i} = 1\,, \quad i = 1, \dots, n\,, \quad \sum_{j=1}^{n+1} \mu_j \tilde{P}_1^{(-\alpha,-\beta)}(t_j) = 0 \tag{27}$$

mit $\tilde{P}_1^{(-\alpha,-\beta)}(t) = t + \alpha - \beta$. Dieses System wird mit der Methode der UV-Reduktion für Cauchy-Matrizen ([9]) schnell aufgelöst. Aus (vgl. (24) und (25))

$$\begin{bmatrix} \left(\dfrac{\mu_j}{t_j - s_i}\right)_{i=1,j=1}^{n,n+1} \\ 1 \quad 2 \quad \dots \quad n+1 \end{bmatrix}^{-1} = \begin{bmatrix} \left(\dfrac{\lambda_i}{t_j - s_i}\right)_{j=1,i=1}^{n+1,n} & \begin{matrix} d \\ d \\ \vdots \\ d \end{matrix} \end{bmatrix}$$

mit $d = \left(\sum_{j=1}^{n+1} \mu_j\right)^{-1}$ folgt

$$\lambda_i \sum_{j=1}^{n+1} \frac{\mu_j}{(t_j - s_i)^2} = 1\,, \quad i = 1, \dots, n\,, \tag{28}$$

zur Berechnung der λ_i, $i = 1, \dots, n$.

Bemerkung 3: Die Gewichte der Quadraturformeln können auch aus den ersten Komponenten der den Eigenwerten s_i und t_j entsprechenden Eigenvektoren ermittelt werden (vgl. z. B. [19]).

4. Lösung der nichtlinearen singulären Integralgleichung

Wir lösen die nichtlineare singuläre Integralgleichung (8) mit Hilfe eines Newton-Quadratur-Verfahrens. Nach der p-ten Iteration ergibt sich

$$u_{p+1}(t) = u_p(t) + \Delta u_p(t)\,, \tag{29}$$

wobei $\Delta u_p(t)$ aus

$$g'[u_p(t) + y_0(t)]\, \Delta u_p(t) - \frac{1}{\pi} \int_{-1}^{1} \frac{\Delta u_p(s)\, ds}{s - t} - D_p = R(t) - \left\{ g([u_p(t) + y_0(t)] - \frac{1}{\pi} \int_{-1}^{1} \frac{u_p(s)}{s - t} ds \right\},$$

$$t \in (-1, 1)\,, \tag{30}$$

mit den Randbedingungen

$$\Delta u_p(-1) = \Delta u_p(1) = 0 \tag{30a}$$

bestimmt wird. Die lineare singuläre Integralgleichung (30) mit einem nichtkonstanten Koeffizienten kann entsprechend der in [11] entwickelten Methode behandelt werden. Wir schreiben hierzu (30) in der Form

$$a(t)\, \tilde{u}(t) - \frac{1}{\pi} \int_{-1}^{1} \frac{\tilde{u}(s)\, ds}{s - t} - D = \psi(t)\,, \quad t \in (-1, 1)\,, \tag{30'}$$

und suchen die Lösung in der Gestalt (vgl. [11], Abschnitt 1)

$$\tilde{u}(t) = \sigma_1(t)\, \zeta(t) \tag{31}$$

mit

$$\sigma_1(t) = (1 + r) \exp\left[\int_{-1}^{1} \frac{\alpha_1(s)}{s - t} ds \right] \Big/ \sqrt{a_1^2 + 1}\,,$$

$$\alpha_1(t) = \tfrac{1}{2}(r_1 t + r_0)\,, \quad r_1 = \alpha(1) + \alpha(-1)\,, \quad r_0 = \alpha(1) - \alpha(-1)\,,$$

$$\alpha(1) = \frac{1}{\pi} \arctan [g'(0)]^{-1}\,, \quad \alpha(-1) = \frac{1}{\pi} \arctan [g'(H)]^{-1}\,, \quad a_1(t) = \cot [\pi \alpha_1(t)]\,.$$

Es folgt

$$1/\sqrt{[a_1(t)]^2 + 1} = \sin [\pi \alpha_1(t)]$$

und

$$\sigma_1(t) = (1 + t) \exp\left[r_1 + (r_1 t + r_0) \frac{1}{2} \ln \frac{1 - t}{1 + t} \right] \sin [\pi \alpha_1(t)]\,. \tag{32}$$

Zur näherungsweisen Lösung von (30) bzw. (30′) benutzen wir das lineare algebraische Gleichungssystem

$$\sum_{i=1}^{n} \left\{ [a(t_j) - a_1(t_j)]\, \sigma_1(t_j)\, l_i(t_j) - \frac{\lambda_i}{s_i - t_j} \right\} \xi_i - D = \psi(t_j)\,, \quad j = 1, \dots, n+1\,. \tag{33}$$

Dabei wird die Gaußsche Quadraturformel bezüglich $\sigma_1(t)$ angewandt, und es bezeichnen $\{l_i(t)\}$ das System der Lagrangeschen Grundpolynome bezüglich der Stützstellen $\{s_i\}$ und ξ_i, $i = 1, \ldots, n$, die gesuchten Näherungswerte für $\zeta(s_i)$. Zur Berechnung der rechten Seiten von (33) verwenden wir

$$u_p(t_j) = \sigma_1(t_j)\, \zeta_p(t_j) = \sum_{i=1}^{n} \zeta_p(s_i)\, \sigma_1(t_j)\, l_i(t_j) \tag{34 a}$$

und

$$-\frac{1}{\pi} \int_{-1}^{1} \frac{u_p(s)\, ds}{s - t_j} = -a_1(t_j)\, u_p(t_j) - \sum_{i=1}^{n} \frac{\lambda_i \zeta_p(s_i)}{s_i - t_j}. \tag{34 b}$$

Bemerkung 4: Die Lösung des Gleichungssystems (33) erfolgte auf der Grundlage des Gaußschen Eliminationsverfahrens mit Spaltenpivot. Die Systemmatrix von (33) läßt sich aber auch als Produkt einer Rang-1-gestörten Löwner-Matrix

$$\begin{bmatrix} \left(\dfrac{h_j - g_i}{c_j - d_i}\right)_{i=1,j=1}^{n,n+1} & \begin{matrix} -1 \\ \vdots \\ -1 \end{matrix} \end{bmatrix}$$

mit einer Diagonalmatrix schreiben, so daß auch hier ein schnelles Auflösungsverfahren auf der Grundlage der UV-Reduktion anwendbar ist ([6], [9]).

Als Startnäherung wird für den ersten k-Wert die Funktion $u_0(t) = 0$ genommen, für jeden weiteren k-Wert die Lösung für den vorangegangenen. Das Iterationsverfahren zur Berechnung von $u_p(t)$ für gegebenes k wird beendet, falls

$$\sum_{i=1}^{n} \lambda_i \xi_i^2 + (D_m - D_{m-1})^2 \leqq \varepsilon^* \tag{35}$$

für ein vorgegebenes $\varepsilon^* > 0$ erfüllt ist. Zur numerischen Realisierung des Verfahrens benötigt man nun lediglich noch einen effektiven Algorithmus zur Berechnung der Stützstellen s_i (d. h. der Nullstellen des bezüglich $\sigma_1(t)$ orthogonalen Polynoms n-ten Grades $P_n(t)$), der Kollokationspunkte t_j (d. h. der Nullstellen des bezüglich

$$[(a_1^2(t) + 1)\, \sigma_1(t)]^{-1} = \mu_1(t)$$

orthogonalen Polynoms $(n + 1)$-ten Grades $Q_{n+1}(t)$) und der Quadraturgewichte λ_i.

Der wesentliche Teil dieses Algorithmus besteht darin, die Rekursionskoeffizienten $d_0, d_1, \ldots, d_n, e_0 e_1, \ldots, e_{n-1}$ für das Polynomsystem $\{Q_j\}$ zu berechnen. Dabei sind die $d_1, \ldots, d_n$ und $e_1, \ldots, e_{n-1}$ gleichzeitig die Rekursionskoeffizienten des Systems $\{P_i\}$, und diese werden aus den Rekursionskoeffizienten $d'_1, \ldots, d'_n, e'_1, \ldots, e'_{n-1}$ der Jacobi-Polynome $\tilde{P}_i^{(\gamma,\delta)} = t^i + \ldots$, $\gamma = 1 - \alpha(-1)$, $\delta = \alpha(1)$, und den Größen

$$m_i = \frac{1}{\pi} \int_{-1}^{1} \sigma_1(t)\, \tilde{P}_i^{(\gamma,\delta)}(t)\, dt\,, \qquad j = 0, 1, \ldots, 2n - 1\,, \tag{36}$$

mittels der Methode der modifizierten Momente (siehe z. B. [7], [8]) berechnet. Dazu verwenden wir die Beziehung

$$m_i = A_1(t\tilde{P}_i^{(\gamma,\delta)})\,(0) \tag{37}$$

mit

$$(A_1 v)\,(t) = a_1(t)\, \sigma_1(t)\, v(t) - \frac{1}{\pi} \int_{-1}^{1} \frac{v(s)}{s - t}\, \sigma_1(s)\, ds\,.$$

Den Operator A_1 kann man auf Polynomen auswerten, wenn man die Entwicklung der stückweise holomorphen Funktion

$$X_1(z) = (1 + z) \exp \int_{-1}^{1} \frac{\alpha_1(t)}{t - z}\, dt$$

im unendlich fernen Punkt kennt,

$$X_1(z) = x_{-1} z + x_0 + x_1 z^{-1} + x_2 z^{-2} + \ldots \tag{38}$$

(vgl. [12], [16]). Es ist nämlich $A_1(t\tilde{P}_i^{(\gamma,\delta)})$ das Polynom $(i + 2)$-ten Grades $R_{i+2}(t)$, für welches

$$\lim_{z \to \infty} [X_1(z)\, z\tilde{P}_i^{(\gamma,\delta)}(z) - R_{i+2}(z)] = 0 \tag{39}$$

gilt. Für die Koeffizienten x_i in (38) gelten die Rekursionsformeln ([12], [16])

$$c_j = \begin{cases} -r_0/j\,, & j \text{ ungerade}\,, \\ -r_1/(j + 1)\,, & j \text{ gerade} \end{cases} \qquad j = 1, 2, \ldots\,; \tag{40 a}$$

$$\alpha_{j1} = c_j; \qquad \alpha_{j,m+1} = \sum_{i=1}^{j} \alpha_{im} c_{j+1-i}\,, \qquad j = 1, 2, \ldots\,, \qquad m = 1, 2, \ldots; \tag{40 b; c}$$

$$y_j = \sum_{m=1}^{j} \frac{1}{m!} \alpha_{j-m+1,m}; \qquad x_{-1} = 1\,, \quad x_0 = 1 + y_1\,, \quad x_j = y_j + y_{j+1}\,, \quad j = 1, 2, \ldots \tag{40 d, e}$$

Die Polynome $\tilde{P}_j^{(\gamma,\delta)}$ sind gegeben durch

$$\tilde{P}_{j+1}^{(\gamma,\delta)}(t) = (t - d'_{j+1})\, \tilde{P}_j^{(\gamma,\delta)}(t) - e_j'^2 \tilde{P}_{j-1}^{(\gamma,\delta)}(t)\,, \qquad j = 0, 1, 2, \ldots,$$

mit $e_0'^2 = 0$ und

$$d'_{j+1} = \frac{\gamma^2 - \delta^2}{(2j + \gamma + \delta)(2j + \gamma + \delta + 2)}\,, \qquad j = 0, 1, 2, \ldots, \tag{41a}$$

$$e_j'^2 = \frac{4j(j + \gamma)(j + \delta)(j + \gamma + \delta)}{(2j + \gamma + \delta - 1)(2j + \gamma + \delta)^2 (2j + \gamma + \delta + 1)}\,, \qquad j = 1, 2, \ldots \tag{41b}$$

Mittels dieser Rekursionskoeffizienten sind die Polynome $\tilde{P}_j^{(\gamma,\delta)}$ in der Form

$$\tilde{P}_j^{(\gamma,\delta)}(t) = t^j + \beta_j^{(j)} t^{j-1} + \ldots + \beta_1^{(j)} \tag{42}$$

darstellbar, und aus (37) und (39) folgt für die Berechnung der modifizierten Momente (36)

$$m_j = x_1\beta_1^{(j)} + \ldots + x_j\beta_j^{(j)} + x_{j+1}\,, \qquad j = 0, 1, \ldots, 2n - 1\,. \tag{43}$$

Schließlich gilt noch

$$d_0 = -x_0\,, \qquad e_0^2 = -x_1\,.$$

Bemerkung 5. Bei der Implementierung des hier in aller Kürze beschriebenen Algorithmus auf einem Bürocomputer A 5120 mittels eines FORTRAN-Programms (unter den Betriebssystem SCPX 1526-V 1.5) hat es sich als erforderlich erwiesen, die Berechnung der Koeffizienten x_j in (38) über die Formeln (40) sowie der Koeffizienten $\beta_i^{(j)}$ in (42) unter Verwendung der Größen (41) mit doppelter Genauigkeit durchzuführen. Gleiches gilt für die Bestimmung der modifizierten Momente nach (43).

Bemerkung 6: Bezüglich einer ausführlichen Konvergenzanalysis der hier vorgestellten Newton-Quadratur-Methode verweisen wir auf [15].

5. Numerische Resultate

Im folgenden geben wir einige auf dem A 5120 erzielte Ergebnisse an und vergleichen diese, soweit möglich, mit in der Literatur angegebenen Resultaten, die auf anderen Methoden beruhen.

Beispiel 1.1 (Lineare Dammkurve): Mit $h = H = 1$, $l = 2$ ergaben sich in Abhängigkeit von der Dimension n der Diskretisierung, der Ordnung m der Tschebyscheff-Quadratur und der Abbruchschranke ε die in Tab. 1 angegebenen Werte für die Durchflußrate q.

Tabelle 1

n	m	ε	q
10	20	10^{-3}	0,36179
40	40	10^{-3}	0,36119
10	60	10^{-4}	0,36110
20	60	10^{-4}	0,36102
40	60	10^{-4}	0,36101

Mittels der auf Variationsungleichungen beruhenden „C. S. Technique" wurde in [24] die Durchflußrate $q = 0{,}36215$ ermittelt. Die nach verschiedenen Verfahren gewonnenen Sickerlinien sind in Abb. 2 dargestellt, wobei die für die Integralgleichungsmethode angegebene Sickerlinie den Parametern der ersten Zeile aus Tab. 1 entspricht.

Bemerkung 7. Für viele Fälle genügt die Ordnung $m = 20$, um die Sickerlinie hinreichend genau zu fixieren. Falls der Parameter k nahe bei 1 liegt, muß m größer gewählt werden, um auch in der Nähe der Endpunkte der Sickerlinie Wertepaare $(x(\tau_j), y(\tau_j))$ zu erhalten. Das ist im Beispiel 1.2 der Fall.

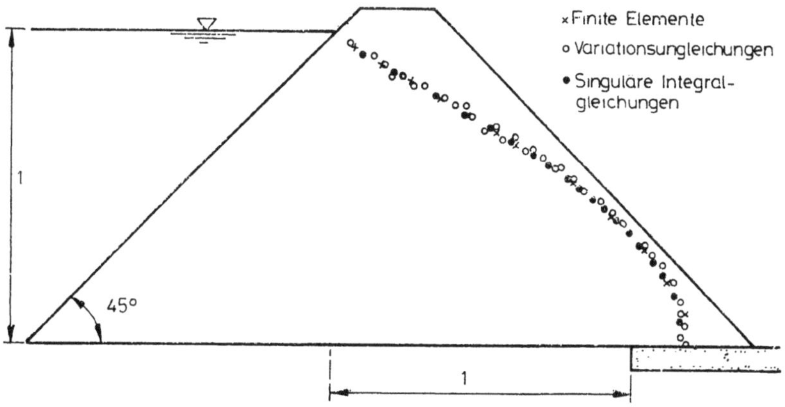

Abb. 2

Beispiel 1.2 (Lineare Dammkurve): Für $h = H = 1$, $l = 3{,}16$, $m = 60$, $\varepsilon = 10^{-4}$ und $n = 10$ bzw. $n = 20$ erhalten wir die Durchflußrate $q = 0{,}20154$ bzw. $q = 0{,}20157$ ([24]: $q = 0{,}20115$). Die entsprechenden Sickerlinien sind in Abb. 3 dargestellt.

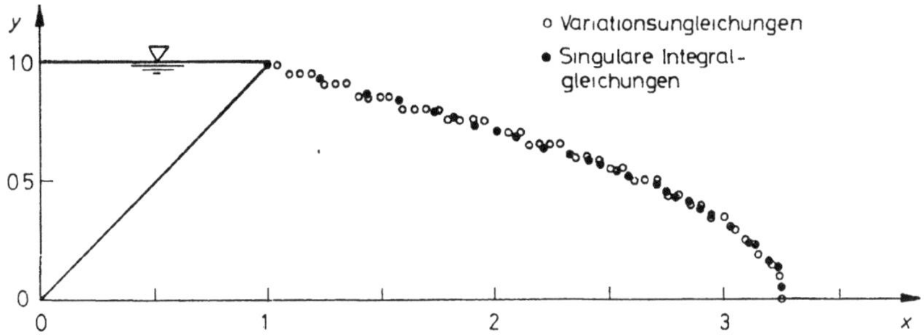

Abb. 3

Beispiel 1.3 (Senkrechter Damm): Wir betrachten den Fall einer senkrechten Dammkurve ($H = 1$, $h = 0$). In [3] (vgl. auch [20]) wird dazu für hinreichend großes l/H eine analytische Näherung für die Sickerlinie in der Form

$$x - l = -(y^2 - q_a^2)/(2q_a) \tag{44a}$$

mit der Durchflußrate

$$q_a = -3l + \sqrt{9l^2 + 3H^2} \tag{44b}$$

angegeben. Wir approximieren diesen Fall durch den Grenzübergang $h \to 0$. In Tab. 2 sind einige Resultate für $h = 10^{-6}$, $\varepsilon = 10^{-5}$ zusammengestellt. Einen Vergleich der Sickerlinien zeigt Abb. 4. Für kleine Verhältnisse l/H sind die Sickerlinien nach (44) praktisch unsinning.

Tabelle 2

l	n	m	q	q_a
2,0	11	60	0,24506	0,24499
2,0	15	60	0,24505	
1,0	10	40	0,46429	0,46410
1,0	15	40	0,46424	
1,0	20	40	0,46423	
0,3	10	60	1,09457	1,05192
0,3	15	60	1,09359	
0,3	20	60	1,09323	
0,1	10	60	1,79480	1,45784
0,1	15	60	1,78429	
0,1	20	60	1,78065	

Beispiel 2 (Gekrümmte Dammkurve): Zur Anwendung der Newton-Quadratur-Methode betrachten wir folgende Beispiele ($h = H = 1$):

Beispiel 2.1: $y = f(x) = (1 - 2b)^{-1} (x - 2b) x$,

Beispiel 2.2: $x = g(y) = [b^2 + (2(1 - b) - y) y]^{1/2} - b$,

Beispiel 2.3: $y = f(x) = b(1/b - 2)^{-1} (e^{ax} - 1)$, $a = 2 \cdot \ln (1/b - 1)$.

Für $l = 2$ und verschiedene Parameterwerte b erhalten wir mit $n = 10$, $m = 40$, $\varepsilon = 10^{-3}$ und $\varepsilon^* = 10^{-4}$ die in Tab. 3 angegebenen Durchflußraten. Wir können feststellen, daß die Durchflußrate um so größer wird, je stärker

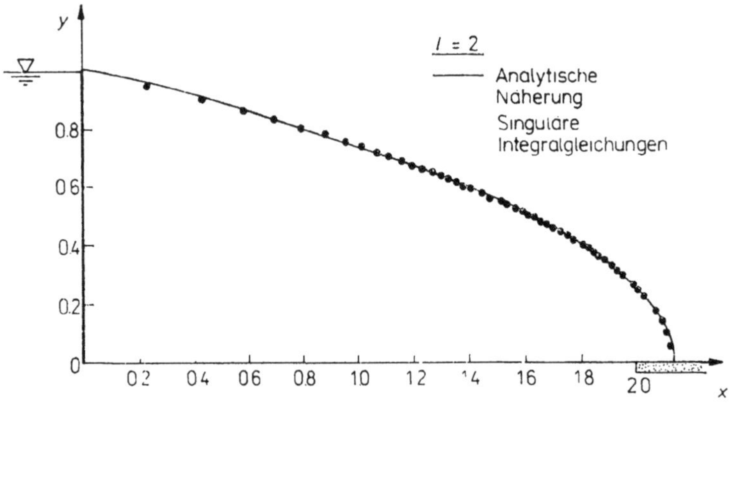

Abb. 4a

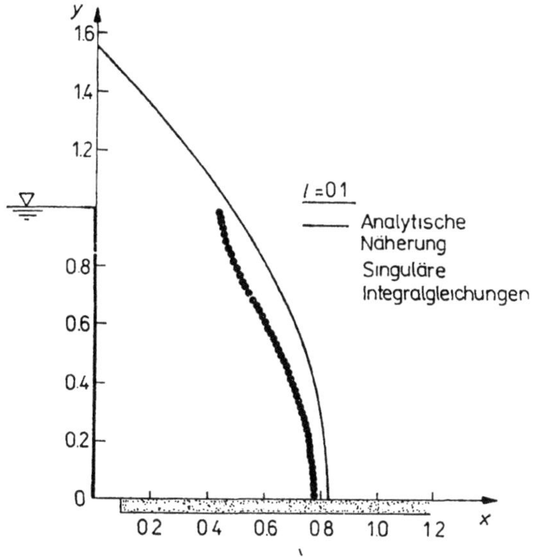

Abb. 4b

die „Durchbiegung" der Dammkurve ist. Allerdings unterscheiden sich die Werte, wie man aus Tab. 3 sieht, nur wenig. Die Sickerlinien sind jeweils für den mittleren Parameter b in den Beispielen 2.1–2.3 sowie im linearen Fall (Beispiel 1.1) zusammen mit den entsprechenden Dammkurven in Abb. 5 dargestellt.

Tabelle 3

Beispiel	b	q
2.1	−5	0,36351
	−0,1	0,38063
	−0,001	0,38483
2.2	1	0,37903
	0,01	0,41346
	0,0001	0,41464
2.3	0,1	0,40805
	0,01	0,43052
	0,001	0,43972

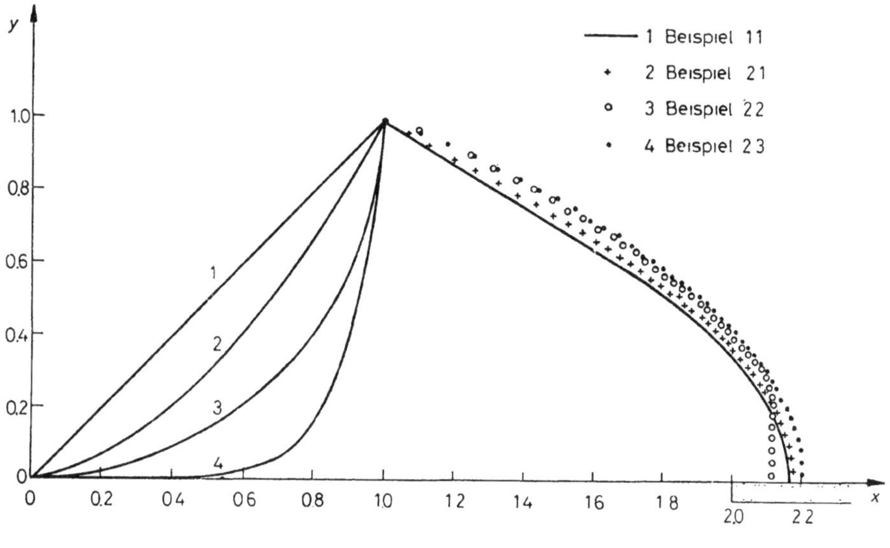

Abb. 5

Literatur

1 BAUMEISTER, J.: Elliptische Variationsungleichungen. GAMM-Mitteilungen (1980) Heft 2, 1–19.

2 BOWDLER, H.; MARTIN, R. S.; REINSCH, C.; WILKINSON, J. H.: The QR- and QL-algorithms for symmetric matrices. Numer. Math. **11** (1968), 293–306.

3 CRANK, J.: Free and moving boundary problems. Clarendon Press, Oxford 1984

4 ELLIOTT, D.: Orthogonal polynomials associated with singular integral equations having a Cauchy kernel. SIAM J. Math. Anal. **13** (1982), 1041–1052.

5 ELLIOTT, D.: The classical collocation method for singular integral equations. SIAM J. Numer. Anal. **19** (1982), 816–832.

6 FLEIG, M.: Schnelle Algorithmen zur Invertierung von Löwner-Matrizen. Jahresarbeit, TU Karl-Marx-Stadt 1987.
7 GAUTSCHI, W.: On the construction of Gaussian quadrature rules from modified moments. Math. Comput. **24** (1970), 245–260.
8 GOLUB, G. H.; WELSCH, J. H.: Calculation of Gauss quadrature rules. Math. Comput. **23** (1969), 221–230.
9 HEINIG, G.; ROST, K.: Algebraic methods for Toeplitz-like matrices and operators. Akademie-Verlag, Berlin, Birkhäuser Verlag, Basel, Boston, Stuttgart 1984.
10 HOFSOMMER, D. J.; RIET, R. P. VAN DE: On the numerical calculation of elliptic integrals of the first and the second kind and the elliptic functions of Jacobi. Numer. Math. **5** (1963), 291–302.
11 JUNGHANNS, P.: Kollokations- und Quadraturformelverfahren zur näherungsweisen Lösung einer Klasse singulärer Integralgleichungen mit einer festen Singularität, Wiss. Z. Techn. Hochsch. Karl-Marx-Stadt **24** (1982) 3, 295–303.
12 JUNGHANNS, P.: Polynomiale Näherungsverfahren für singuläre Integralgleichungen auf beschränkten Intervallen. Dissertation B, TH Karl-Marx-Stadt 1983.
13 JUNGHANNS, P.: Some remarks on the zero distribution of pairs of polynomials associated with singular integral operators. A convergence theorem for the quadrature method. Wiss. Z. d. Techn. Hochsch. Karl-Marx-Stadt **27** (1985) 1, 88–93.
14 JUNGHANNS, P.: Effective solution of systems of algebraic equations occurring in the approximate solution of singular integral equations by means of the method of quadrature formulae. Wiss. Z. Techn. Hochsch. Karl-Marx-Stadt **27** (1985) 1, 94–96.
15 JUNGHANNS, P.: Numerical analysis of Newton-projection-methods for nonlinear singular integral equations. In Vorbereitung.
16 JUNGHANNS, P.; SILBERMANN, B.: Zur Theorie der Näherungsverfahren für singuläre Integralgleichungen auf Intervallen. Math. Nachr. **103** (1981), 199–244.
17 JUNGHANNS, P.; SILBERMANN, B.: Numerical analysis for one-dimensional Cauchy-type singular integral equations. In: Probleme und Methoden der Mathematischen Physik, Teubner-Texte zur Mathematik, Leipzig 1984, pp. 122–129.
18 JUNGHANNS, P.; SILBERMANN, B.: The numerical treatment of singular integral equations by means of polynomial approximations. Part I. Preprint, P-MATH-35/86, Akad. d. Wiss. d. DDR, Berlin 1986. Part II. In Vorbereitung.
19 KAUTSKY, J.; ELHAY: Calculation of the weights of interpolatory quadratures. Numer. Math. **40** (1982), 407–422.
20 KOZENY, J.: Grundwasserbewegung bei freiem Spiegel, Fluß und Kanalversickerung. Wasserkraft und Wasserwirtschaft **3** (1931), 28–31.
21 OESTREICH, D.: Zum Staudammproblem mit Drainage. Z. angew. Math. Mech. **67** (1987) 7, 293–300.
22 OESTREICH, D.: Ein Problem der elektrochemischen Bearbeitung mit gekrümmter Katode. Erscheint in Demonstratio Math.
23 Полубаринова-Кочина, П. Я.: Теория движения грунтовых вод. Издательство "Наука", Москва 1977.
24 SLOSS, J. M.; BRUCH, J. C., Jr.: Free-surface seepage problem. J. Engineering Mech. Div., ASCE **104** (1978), No. EM 5, Proc. Paper 14058, 1099–1111.
25 WHEELER, J. C.: Modified moments and Gaussian quadratures. Rocky Mountain J. Math. **4** (1974) 2, 287–296.
26 WOLFERSDORF, L. v.: Zur Regularisierung von Cauchy-Problemen für analytische und verallgemeinerte analytische Funktionen (Studie). Bergakademie Freiberg 1984.

Eingegangen: 6. Januar 1988.

Anschriften: Dr. PETER JUNGHANNS, Technische Universität Karl-Marx-Stadt, Sektion Mathematik, PSF 964, DDR-9010 Karl-Marx-Stadt; Dr. DIETER OESTREICH, Bergakademie Freiberg, Sektion Mathematik, B.-v.-Cotta-Str. 2, DDR-9200 Freiberg

ZAMM Z. angew. Math. Mech. 69 (1989) 2, 93 – 96

CARL, S.

A Comparison Result for Nonlinear Elliptic Systems Having Nonmonotone Nonlinearities

Es wird ein Vergleichssatz für Randwertprobleme nichtlinearer elliptischer Systeme bewiesen. Gewöhnlich sind dafür Monotoniebedingungen an die nichtlinearen rechten Seiten erforderlich. In dieser Arbeit soll ein Vergleichssatz angegeben werden, der keinerlei Monotonieforderungen an die rechten Seiten stellt.

A comparison theorem for boundary value problems of nonlinear elliptic systems is proved. Usually, such comparison theorems require certain monotonicity conditions on the nonlinear right-hand sides. In this paper a comparison theorem is given that does not impose any monotonicity restrictions on the right-hand sides.

Доказывается теорема сравнения для краевых задач нелинейных эллиптических систем. Обычно такая теорема требует некоторых свойств монотонности для нелинейных правых частей. В этой работе дана теорема сравнения, для которой не нужны свойства монотонности.

1. Introduction

Comparison principles have been proved to be powerful tools in the theory of differential equations with regard to various qualitative properties of the solutions such as a priori bounds, uniqueness or invariant set results (cf. [1 4, 7, 8, 10, 11, 12]). Usually papers dealing with comparison theorems impose monotonicity restrictions on the nonlinear right-hand sides (cf. [3, 5]).

The main objective of this paper is to establish a comparison theorem for boundary value problems of nonlinear elliptic systems even when the right-hand sides do not possess any monotonicity properties.

Recently an approach in this direction has been given by MCNABB in [9], but his method is different from ours. In addition, in our treatment the smoothness assumptions on the coefficients and the right-hand sides are weakened to a wide extent.

Finally, the comparison theorem to be proved in this paper can be used to establish a monotone iterative scheme that yields a monotone approximation of the solution even in the case of nonmonotone right-hand sides.

2. Notations and definitions

This paper is concerned with the following elliptic boundary value problem (BVP for short)

$$-L_k u_k = f_k(x, u) \quad \text{in } \Omega\,, \qquad B_{k\delta} u_k = 0 \quad \text{on } \partial\Omega \tag{1), (2}$$

for $k = 1, \dots, n$, where $u\colon \Omega \to R^n$ and Ω is a bounded domain in R^N having a regular (Lipschitz continuous) boundary $\partial\Omega$. Without loss of generality homogeneous boundary values can be taken in (2).

For $\delta \in \{0, 1\}$ the boundary operators $B_{k\delta}$ are defined

$$B_{k0} u_k = u_k \quad \text{and} \quad B_{k1} u_k = (\partial u_k/\partial v^k) + b^k(x)\, u_k\,,$$

where $\partial/\partial v^k$ denotes the outward conormal derivative on $\partial\Omega$ with respect to the strictly uniformly elliptic differential operator L_k given by

$$L_k = \frac{\partial}{\partial x_i}\left(a_{ij}^k(x)\,\frac{\partial}{\partial x_j}\right), \qquad k = 1, \dots, n\,.$$

(Throughout this paper we use the summation convention.)

Let the coefficients a_{ij}^k and b^k be real, bounded and measurable in their respective domains Ω and $\partial\Omega$, and let b^k be nonnegative. The matrices $A_k = (a_{ij}^k)$ are assumed to be symmetric. As usual, we denote by $H^1(\Omega)$ the Sobolev space of all square integrable functions having square integrable weak derivatives of first order, that is

$$H^1(\Omega) = \{u \in L_2(\Omega) \mid \partial u/\partial x_i \in L_2(\Omega)\}\,.$$

The subspace of all functions with homogeneous generalized boundary values is denoted by $H_0^1(\Omega)$. The corresponding Sobolev spaces of vector functions $u\colon \Omega \to R^n$ are denoted by $H^1(\Omega, R^n)$ and $H_0^1(\Omega, R^n)$, respectively.

According to the boundary operators $B_{k\delta}$ we introduce the bilinear forms $l_{k\delta}$ by

$$l_{k\delta}(u, v) = \int_\Omega a_{ij}^k \frac{\partial u}{\partial x_j}\frac{\partial v}{\partial x_i}\,dx + \delta \int_{\partial\Omega} b^k u v\, ds\,.$$

Definition: The vector function $u \in H_0^1(\Omega, R^n)$ $\big(H^1(\Omega, R^n)\big)$ is called a *weak solution* of the BVP (1), (2) if the integral relations

$$l_{k\delta}(u_k, \chi) = \int_\Omega f_k(x, u)\,\chi\, dx\,, \qquad k = 1, 2, \dots, n\,,$$

hold for all $\chi \in H_0^1(\Omega)$ $\big(H^1(\Omega)\big)$ and $\delta = 0$ $(\delta = 1)$. □

Let us denote by $[u, v]$ the order interval in $L_2(\Omega, R^n)$, that is

$$[u, v] = \{z \in L_2(\Omega, R^n) \mid u_k(x) \leqq z_k(x) \leqq v_k(x)\,,\ \text{for } k = 1, \dots, n \text{ and almost everywhere in } \Omega\}\,.$$

Further, the vector $[u]_k \in R^{n-1}$ is defined by

$$[u]_k = (u_1, \dots, u_{k-1}, u_{k+1}, \dots, u_n)\,.$$

Let $M = (M_{ij})$ be an (n,n)-matrix. Then $M = D^M + M^0$, where D^M is diagonal with $D^M_{ii} = M_{ii}$, $M^0_{ii} = 0$ and $M^0_{ij} = M_{ij}$ if $i \neq j$.

The vector $[M]_k \in R^{n-1}$ of elements of the matrix M is given by

$$[M]_k = (M_{k1}, \dots, M_{k,k-1}, M_{k,k+1}, \dots, M_{kn})\,.$$

Consider the eigenvalue problems

$$L_k\varphi_k + \lambda_k\varphi_k = 0 \quad \text{in } \Omega\,, \qquad B_{k\delta}\varphi_k = 0 \quad \text{on } \partial\Omega \tag{3), (4}$$

for $k = 1, \dots, n$, then by D^λ we denote the diagonal matrix the elements of which are the least eigenvalues of (3), (4).

Using the notations introduced above the right-hand sides f_k of equation (1) can be rewritten in the form $f_k(x, u) = f_k(x, u_k, [u]_k)$. Finally, we introduce functions F_k defined by

$$F_k(x, u_k, [v]_k) = f_k(x, u_k, [v]_k) - [M]_k\,[v]_k\,.$$

In what follows $||\cdot||$ stands for the norm in $L_2(\Omega)$.

3. Comparison theorem

Let I be some interval in $L_2(\Omega, R^n)$ then we make the following hypotheses:

(H 1): Let the right-hand sides f_k satisfy a uniform Lipschitz condition of the form

$$|f_k(\cdot, u) - f_k(\cdot, v)| \leqq \gamma_{ki}|u_i - v_i| \quad \text{for} \quad u, v \in I\,, \qquad k = 1, \dots, n\,.$$

(H 2): Let the matrix $M = (M_{ij})$, $M_{ij} \geqq 0$, to be chosen in such a way that the functions F_k are quasi-monotone nonincreasing in I, and let there exist functions $\bar{u}$, $\underline{u} \in H^1(\Omega, R^n)$ of the interval I satisfying the following inequalities:

$$-L_k\bar{u}_k - [M]_k[\bar{u}]_k \geqq F_k(x, \bar{u}_k, [\underline{u}]_k) \quad \text{in } \Omega\,, \qquad B_{k\delta}\bar{u}_k \geqq 0 \quad \text{on } \partial\Omega\,, \tag{5), (6}$$

and

$$-L_k\underline{u}_k - [M]_k\,[\underline{u}]_k \leqq F_k(x, \underline{u}_k, [\bar{u}]_k) \quad \text{in } \Omega\,, \qquad B_{k\delta}\underline{u}_k \leqq 0 \quad \text{on } \partial\Omega\,, \tag{7), (8}$$

where "$[\cdot]_k\,[\cdot]_k$" means the scalar product in R^{n-1}.

Remarks: i) The quasi-monotonicity of the functions F_k can always be obtained due to hypothesis (H 1). ii) Of course, the differential inequalities (5)—(8) have to be understood in an appropriate weak sense whose precise meaning will be given below.

Now we are able to prove our main Theorem.

Theorem: *Let the hypotheses* (H 1) *and* (H 2) *be satisfied for a pair of vector functions* $\bar{u}$, $\underline{u} \in H^1(\Omega, R^n)$ *with respect to the order interval* $I = [\inf(\bar{u}, \underline{u}), \sup(\bar{u}, \underline{u})]$. *Then the inequality* $\bar{u} \geqq \underline{u}$ *holds, provided the matrix* $D^\lambda - \gamma - 2M^0$ *is positive definite.*

Proof: The proof will be given only for the Dirichlet boundary condition, i.e. $\delta = 0$, because the proof for the case $\delta = 1$ follows essentially the same arguments.

Due to hypothesis (H 2) the functions $\bar{u}$, $\underline{u}$ satisfy inequalities (5)—(8) in the following weak sense:

$$l_{k0}(\bar{u}_k, \chi) - \int_\Omega [M]_k\,[\bar{u}]_k\,\chi\,\mathrm{d}x \geqq \int_\Omega F_k(x, \bar{u}_k, [\underline{u}]_k)\,\chi\,\mathrm{d}x\,,$$

$$l_{k0}(\underline{u}_k, \chi) - \int_\Omega [M]_k\,[\underline{u}]_k\,\chi\,\mathrm{d}x \leqq \int_\Omega F_k(x, \underline{u}_k, [\bar{u}]_k)\,\chi\,\mathrm{d}x$$

for all nonnegative test functions $\chi \in H^1_0(\Omega)$, and $\bar{u}_k(x) \geqq 0 \geqq \underline{u}_k(x)$ on $\partial\Omega$. By subtracting corresponding inequalities from each other we get the following inequalities for the auxiliary function $w_k = \underline{u}_k - \bar{u}_k$:

$$l_{k0}(w_k, \chi) - \int_\Omega [M]_k\,[w]_k\,\chi\,\mathrm{d}x \leqq \int_\Omega \big(F_k(x, \underline{u}_k, [\bar{u}]_k) - F_k(x, \bar{u}_k, [\underline{u}]_k)\big)\,\chi\,\mathrm{d}x\,, \tag{9}$$

and $w_k \leqq 0$ on $\partial\Omega$.

Now, as a special test function we choose $\chi = w_k^+$, where w_k^+ is defined by $w_k^+ = \max_\Omega (w_k, 0)$. In [6, Lemma 7.6.] it is shown that $w_k^+ \in H^1(\Omega)$ and because of $w_k \leqq 0$ on $\partial\Omega$ it must even be an element of $H^1_0(\Omega)$. Thus w_k^+ is an admissible test function, since obviously $w_k^+ \geqq 0$. Taking into account the quasi-monotonicity and the Lipschitz continuity of the functions F_k we can estimate the right-hand side of inequality (9) as follows:

$$\int_\Omega \big(F_k(x, \underline{u}_k, [\bar{u}]_k) - F_k(x, \bar{u}_k, [\underline{u}])_k\big)\,w_k^+\,\mathrm{d}x =$$

$$= \int_\Omega \big(F_k(x, \underline{u}_k, [\bar{u}]_k) - F_k(x, \bar{u}_k, [\bar{u}]_k)\big)\,w_k^+\,\mathrm{d}x + \int_\Omega \big(F_k(x, \bar{u}_k, [\bar{u}]_k) - F_k(x, \bar{u}_k, [\underline{u}]_k)\big)\,w_k^+\,\mathrm{d}x \leqq$$

$$\leqq \int_\Omega \{\gamma_{kk}|\underline{u}_k - \bar{u}_k|\,w_k^+ + [\gamma + M]_k\,[(\underline{u} - \bar{u})^+]_k\,w_k^+\}\,\mathrm{d}x \leqq \int_\Omega \{\gamma_{kk}(w_k^+)^2 + [\gamma + M]_k\,[w^+]_k\,w_k^+\}\,\mathrm{d}x\,. \tag{10}$$

Furthermore, it is easily seen that

$$\int_\Omega [M]_k [w]_k w_k^+ \, dx \leqq \int_\Omega [M]_k [w^+]_k w_k^+ \, dx , \qquad l_{k0}(w_k, w_k^+) = l_{k0}(w_k^+, w_k^+) \tag{11), (12}$$

(see Lemma 7.6. of [6]).

It is an immediate consequence of the variational formulation of the eigenvalue problem (3), (4) (see [6, Chap. 8.12.]) that the least eigenvalue λ_k satisfies the inequality

$$\lambda_k \|w_k^+\|^2 \leqq l_{k0}(w_k^+, w_k^+) . \tag{13}$$

With (10)—(13) from (9) we get the estimate

$$\lambda_k \|w_k^+\|^2 \leqq \gamma_{kk} \|w_k^+\|^2 + \int_\Omega [2M + \gamma]_k [w^+]_k w_k^+ \, dx . \tag{14}$$

Applying Schwartz' inequality to the second term on the right-hand side of (14) we get the inequality

$$(\lambda_k - \gamma_{kk}) \|w_k^+\|^2 - (2M_{kj} + \gamma_{kj}) \|w_j^+\| \, \|w_k^+\| \leqq 0$$

for all $k = 1, \ldots, n, j \neq k$. Since the matrix $D^\lambda - \gamma - 2M^0$ is assumed to be positive definite, the last inequality implies that $\|w_i^+\|$ must be zero for all $i = 1, \ldots, n$, and hence the assertion of the Theorem.

Corollary: *Let the hypotheses of the Theorem be satisfied. Then there is at most one solution u of the* BVP (1), (2) *within the interval I.*

Proof: We consider again only the case $\delta = 0$ since the case $\delta = 1$ can be treated analogously.

Let be given two solutions u^1 and u^2. Then $w = u^1 - u^2$ satisfies the relation

$$l_{k0}(w_k, \chi) = \int_\Omega (f_k(x, u^1) - f_k(x, u^2)) \chi \, dx$$

for all $\chi \in H_0^1(\Omega)$, and it holds $w_k(x) = 0$ on $\partial\Omega$. As a special test function we choose $\chi = w_k$. An estimate using the least eigenvalue of (3), (4) and the Lipschitz continuity of the f_k yields $\lambda_k \|w_k\|^2 \leqq \gamma_{kj} \|w_j\| \, \|w_k\|$, or

$$\lambda_k \|w_k\|^2 - \gamma_{kj} \|w_j\| \, \|w_k\| \leqq 0 . \tag{15}$$

The matrix $D^\lambda - \gamma - 2M^0$ is assumed to be positive definite. Since the elements of the matrix M^0 are nonnegative it follows from (15) that $\|w_k\|$ must be zero for all $k = 1, \ldots, n$. This can be seen as follows. Let $\xi \in R^n$, then $\xi(D^\lambda - \gamma - 2M^0) \xi^T \geqq 0$. If we take, in particular, vectors $\xi^+ \in R^n$ of nonnegative elements then we get the estimate

$$0 \leqq \xi^+(D^\lambda - \gamma - 2M^0) \xi^{+T} \leqq \xi^+(D^\lambda - \gamma) \xi^{+T} . \tag{16}$$

Now, for the special nonnegative vector $\xi^+ = (\|w_1\|, \ldots, \|w_n\|)$ we obtain from (15) $\xi^+(D^\lambda - \gamma) \xi^{+T} \leqq 0$, and hence by (16) follows $\xi^+(D^\lambda - \gamma - 2M^0) \xi^{+T} = 0$. The latter can, however, only be satisfied if $\xi^+ = 0$, that means $\|w_k\| = 0$ for all $k = 1, \ldots, n$, and thus the assertion of the Corollary.

Hint: If the hypothesis (H 1) is satisfied globally then the assertion of the Corollary also holds globally. Notice, hypothesis (H 2) is not needed for proving the Corollary.

Remark: The Theorem enables us to establish a monotone iterative scheme for the BVP (1), (2) that yields sequences $\{\overline{u}^i\}$ and $\{\underline{u}^i\}$ of vector functions converging monotonically in $L_2(\Omega, R^n)$ to the uniquely defined solution from above and below, respectively. This can be shown in a similar way as it was done in [2] for initial-boundary value problems of nonlinear parabolic systems.

4. Example

Here we give an example showing that all the hypotheses of the Theorem can be satisfied even though the nonlinear right-hand sides have properties of nonmonotonicity. For this purpose let us consider the following BVP:

$$-\frac{d^2u_1}{dx^2} = \sin px \sin (u_1 - r) \cos u_2 , \qquad -\frac{d^2u_2}{dx^2} = \sin qx \cos u_1 \sin (u_2 - r) , \tag{17}$$

$$x \in \Omega = (0, 1) \quad \text{and} \quad u_k(l) = 0 , \qquad k = 1, 2 , \quad \text{and} \quad l = 0, 1 .$$

The constants p, q and r may be arbitrary, but different from zero. The least eigenvalues of the corresponding eigenvalue problems (3), (4) are $\lambda_1 = \lambda_2 = \pi^2$. Obviously the right-hand sides of (17) are globally Lipschitz continuous. For the matrices M^0 and γ we obtain

$$M^0 = \begin{pmatrix} 0 & 1 \\ 1 & 0 \end{pmatrix} \quad \text{and} \quad \gamma = \begin{pmatrix} 1 & 1 \\ 1 & 1 \end{pmatrix} .$$

One easily verifies that the matrix $D^\lambda - 2M^0 - \gamma$ is positive definite. In order to satisfy hypothesis (H 2) we have to find a pair of vector functions $\underline{u}, \overline{u} \in H^1(\Omega, R^2)$ such that the following inequalities are fulfilled:

$$-\frac{d^2\overline{u}_1}{dx^2} - \overline{u}_2 \geqq \sin px \sin (\overline{u}_1 - r) \cos \underline{u}_2 - \underline{u}_2 , \qquad -\frac{d^2\overline{u}_2}{dx^2} - \overline{u}_1 \geqq \sin qx \cos \underline{u}_1 \sin (\overline{u}_2 - r) - \underline{u}_1 ,$$

$$-\frac{d^2\underline{u}_1}{dx^2} - \underline{u}_2 \leqq \sin px \sin (\underline{u}_1 - r) \cos \overline{u}_2 - \overline{u}_2 , \qquad -\frac{d^2\underline{u}_2}{dx^2} - \underline{u}_1 \leqq \sin qx \cos \overline{u}_1 \sin (\underline{u}_2 - r) - \overline{u}_1 ,$$

and $\underline{u}_k(l) \leqq 0 \leqq \overline{u}_k(l)$, $k = 1, 2$, $l = 0, 1$.

To get a solution of the inequalities above we set $\bar{u}_1 = \bar{u}_2 = A(x^2 - x)$ and $\underline{u}_1 = \underline{u}_2 = B(x^2 - x)$, where the constants A and B have to be chosen appropriately. Carrying out some elementary estimates we obtain the following conditions for the coefficients A and B: $-7A - B \geqq 4$ and $7B + A \geqq 4$. One of the solutions of the last inequalities is, for example, $A = -4$, $B = 24$. Thus all the hypotheses of the Theorem are satisfied.

By the remark made at the end of Chapter 3 and by the Corollary it follows that BVP (17) possesses a uniquely defined solution $u = (u_1, u_2)$ satisfying the inequality

$$24(x^2 - x)\,(1, 1) \leqq (u_1, u_2) \leqq -4(x^2 - x)\,(1, 1)$$

and that the solution $u = (u_1, u_2)$ can be constructed by means of a monotone iterative scheme.

References

1 AMAN, H.: Invariant sets and existence theorems for semilinear parabolic and elliptic systems. J. Math. Anal. Appl. **65** (1978), 432–467.
2 CARL, S.: A monotone iterative scheme for nonlinear reaction-diffusion systems having non-monotone reaction terms. J. Math. Anal. Appl. (accepted for publication).
3 CHANDRA, J.; LEELA, S.: Comparison principle and theory of nonlinear boundary value problems. In: Nonlinear phenomena in mathematical sciences. Proc. int. Conf. Arlington, Tex. 1980, Academic Press, New York-San Francisco-London 1982, pp. 241 to 248.
4 CONWAY, E. D.; SMOLLER, J. A.: A comparison technique for systems of reaction-diffusion equations. Commun. in PDE's **2** (1977), 679 –697.
5 DHAIGUDE, D. B.; KASTURE, D. Y.: Comparison theorems for nonlinear elliptic systems of second order with applications. J. Math. Anal. Appl. **112** (1985), 178–189.
6 GILBARG, D.; TRUDINGER, N. S.: Elliptic partial differential equations of second order. 2. ed., Springer-Verlag, Berlin-Heidelberg-New York-Tokyo 1983.
7 KAWOHL, B.: On maximum principles and Liouville theorems for quasilinear elliptic equations and systems. Comment. Univ. Carol. **24** (1983), 647–655.
8 LAKSHMIKANTHAM, V.; LEELA, S.: Differential and integral inequalities. I, II. Academic Press, New York 1969.
9 McNABB, A.: Comparison theorems for differential equations. J. Math. Anal. Appl. **119** (1986), 417–428.
10 PROTTER, M. H.; WEINBERGER, H. F.: Maximum principles in differential equations. Springer-Verlag, New York-Berlin-Heidelberg-Tokyo 1984.
11 REDHEFFER, R.; WALTER, W.: Invariant sets for systems of partial differential equations. I: Parabolic equations. Arch. Rat. Mech. Anal. **67** (1978), 41–52.
12 WEINBERGER, H.: Invariant sets for weakly coupled parabolic and elliptic systems. Rend. Mat. Univ. Roma 8 (1975), 295–310.

Received January 18, 1988.

Address: Dr. SIEGFRIED CARL, Sektion Mathematik der Technischen Hochschule „Carl Schorlemmer" Leuna-Merseburg, Otto-Nuschke-Straße, DDR-4200 Merseburg

SHORT COMMUNICATIONS

ZAMM · Z. angew. Math. Mech. 69 (1989) 2, 97 – 100

SRINIVASAN, V.; KANDASWAMY, P.; DEBNATH, L.

Propagation of Alfvén Waves in an Isothermal Atmosphere Including the Effects of Displacement Currents

1. Introduction

The atmosphere above 80 km exhibits a wide range of dynamical phenomena. These are largely hydrodynamic in nature, and their sources of energy are likely to lie principally at lower levels. The regions are ionized, however, and hydromagnetically controlled motions also exist. They are generated from above as well as below. The upper atmosphere is indeed the site of substantial motions. The motions of the upper atmosphere are responsible through a dynamo action for the geomagnetic variations that are observed at ground level.

The atmosphere is structured by the pull of gravity in such a way that its density decreases continuously with increase in height. It supports gravity waves that are analogous to those of the oceans.

The propagation of hydromagnetic waves in rotating and non-rotating diffusionless system is relevant to some observed geomagnetic and astrophysical phenomena. One of the important results emerging from the study of hydromagnetic inertial waves in stratified fluids is the existence of critical levels. In systems in which the background state on which the waves ride varies in one independent variable, z (say), critical levels appear as regular singular points of the governing second-order partial differential equation.

The behaviour of waves near the critical levels under various circumstances has been investigated by several authors including BOOKER and BRETHERTON (1967), RUDRAIAH and VENKATACHALAPPA (1972), ACHESON (1972), MCKENZIE (1973), ELTAYEB and KANDASWAMY (1979).

The displacement current is involved in a number of indirect ways, particularly through the electromagnetic theory of light and the electromagnetic mass of the electron. The effects produced by the displacement current seem to be significant in some problems in which conductors of high resistance and in rapidly changing fields. Although the effect of the displacement current in metals is insignificant, but it is very large in a wave problem dealing with oscillating fields of which the frequency is comparable with that of light waves.

In a very rapidly varying field, in which the rate of change of electric displacement with the time is so great that the displacement current is large compared with the conduction current, or so that magnetic forces produced by it are comparable with the electric forces, the existence of electromagnetic waves is possible only because of the presence of the displacement current.

On the other hand, when the displacement current is taken into account, the medium rarefies with altitude, the Alfvén waves resemble more and more electromagnetic waves in a vacuum. At an infinitely high altitude where density is almost zero, they exhibit a travelling behaviour.

Motivated by the above discussion, the effect of displacement currents on the propagation properties of Alfvén gravity waves in an isothermal atmosphere is investigated in this paper. It is shown that the number of singularities of the wave equation increases when the displacement current is included. So the effect of the displacement is to increase the number of critical layers and then reduces the gross attenuation of waves across the critical levels.

2. Derivation of governing wave equation

The unsteady equations of motion for an inviscid, incompressible, perfectly conducting fluid permeated by a uniform magnetic field are given by

$$\varrho\left[\frac{\partial \underline{u}}{\partial t} + \underline{u}\cdot\nabla\underline{u}\right] = -\nabla p + \varrho\underline{g} + \frac{1}{c}(\underline{j}\times\underline{B}) \tag{2.1}$$

$$\frac{\partial \underline{B}}{\partial t} = \nabla\times(\underline{u}\times\underline{B}) \tag{2.2}$$

$$\nabla\cdot\underline{u} = 0 \tag{2.3}$$

$$\nabla\cdot\underline{B} = 0 \tag{2.4}$$

$$\frac{\partial\varrho}{\partial t} + \underline{u}\cdot\nabla\varrho = 0 \tag{2.5}$$

where ϱ, p, $\underline{u}$ represent the fluid density, pressure and velocity fields respectively, $\underline{B}$ and $\underline{g}$ are the magnetic and gravitational fields respectively, $\underline{j}$ is the current density and c is the speed of light.

When the displacement current is taken into account, Ampere's law for perfectly conducting fluid is

$$\nabla\times\underline{B} = \frac{4\pi}{c}\underline{j} - \frac{1}{c^2}\frac{\partial}{\partial t}(\underline{u}\times\underline{B}). \tag{2.6}$$

Substitution of (2.6) into (2.1) gives

$$\varrho\frac{\mathrm{D}\underline{u}}{\mathrm{D}t} = -\nabla p + \varrho\underline{g} + \frac{1}{\mu}\left[\nabla\times\underline{B} + \frac{1}{c^2}\frac{\partial}{\partial t}(\underline{u}\times\underline{B})\right]\times\underline{B}, \tag{2.7}$$

where μ denotes the permeability.

We select a set of Cartesian coordinates (x, y, z) relative to the basic equilibrium state:

$$\underline{u} = [U(z), 0, 0], \qquad \underline{B} = [B(z), 0, 0] \tag{2.8ab}$$

$$\varrho = \varrho_0(z), \qquad \underline{g} = (0, 0, -g). \tag{2.9ab}$$

This basic equilibrium state provides the relation

$$\frac{\mathrm{d}p_0}{\mathrm{d}z} = -\left(\varrho_0 g + \frac{B}{\mu}\frac{\mathrm{d}B}{\mathrm{d}z}\right). \tag{2.10}$$

If we slightly perturb the system, then we have

$$\underline{u} = [U(z) + u_x', u_y', u_z'], \qquad \underline{B} = [B(z) + b_x', b_y', b_z'], \tag{2.11ab}$$

$$p = p_0 + p', \qquad \varrho = \varrho_0 + \varrho' \tag{2.12ab}$$

where the quantities with primes represent the perturbed quantities which are assumed to be small compared to the basic unperturbed quantities.

We substitute (2.11ab)–(2.12ab) into (2.2) –(2.5) and (2.7) and linearize then to obtain the linearized equations governing the motion in the form, dropping the primes,

$$\varrho_0\left[\frac{\partial u_x}{\partial t} + U\frac{\partial u_x}{\partial x} + u_z\frac{\mathrm{d}U}{\mathrm{d}z}\right] = -\frac{\partial p}{\partial x} + \frac{b_z}{\mu}\frac{\mathrm{d}B}{\mathrm{d}z} \tag{2.13}$$

$$\varrho_0\left[\frac{\partial u_y}{\partial t} + U\frac{\partial u_y}{\partial x}\right] = -\frac{\partial p}{\partial y} + \frac{B}{\mu}\left[\frac{\partial b_y}{\partial x} - \frac{\partial b_x}{\partial y} + \frac{U}{c^2}\frac{\partial b_y}{\partial t} - \frac{B}{c^2}\frac{\partial u_y}{\partial t}\right] \tag{2.14}$$

$$\varrho_0\left[\frac{\partial u_z}{\partial t} + U\frac{\partial u_z}{\partial x}\right] = -\frac{\partial p}{\partial z} - \varrho g - \frac{1}{\mu}\left[\frac{\partial}{\partial z}(Bb_x) - B\frac{\partial b_z}{\partial x} - \frac{UB}{c^2}\frac{\partial b_z}{\partial t} + \frac{B^2}{c^2}\frac{\partial u_z}{\partial t}\right] \tag{2.15}$$

$$\frac{\partial\varrho}{\partial t} + U\frac{\partial\varrho}{\partial x} + u_z\frac{\partial\varrho_0}{\partial z} = 0 \tag{2.16}$$

$$\frac{\partial u_x}{\partial x} + \frac{\partial u_y}{\partial y} + \frac{\partial u_z}{\partial z} = 0 \tag{2.17}$$

$$\frac{\partial b_x}{\partial x} + \frac{\partial b_y}{\partial y} + \frac{\partial b_z}{\partial z} = 0 \tag{2.18}$$

$$\frac{\partial b_x}{\partial t} = U\left[\frac{\partial b_y}{\partial y} + \frac{\partial b_z}{\partial z}\right] - B\left[\frac{\partial u_y}{\partial y} + \frac{\partial u_z}{\partial z}\right] + b_z\frac{\mathrm{d}U}{\mathrm{d}z} - u_z\frac{\mathrm{d}B}{\mathrm{d}z} \tag{2.19}$$

$$\frac{\partial b_y}{\partial t} = -\left[U \frac{\partial b_y}{\partial x} - B \frac{\partial u_y}{\partial x}\right] \tag{2.20}$$

$$\frac{\partial b_z}{\partial t} = B \frac{\partial u_z}{\partial x} - U \frac{\partial b_z}{\partial x} . \tag{2.21}$$

The linearized equations of motion (2.13)–(2.21) admit solutions in which all perturbed quantities φ' may be written as

$$\varphi' = \mathrm{Re}\,[\varphi(z) \exp i(kx + ly - \sigma t)] \tag{2.22}$$

where (k, l) is the wave number vector and σ is the frequency.

Substituting this perturbed quantity into (2.13)–(2.21) and eliminating all perturbation variables in favour of the z-component of the velocity component u_z, we obtain

$$\frac{d^2u_z}{dz^2} + \left[\frac{D'}{D} - \beta - \frac{F'}{F}\right]\frac{du_z}{dz} + \\ + \left[\frac{k}{\omega}\frac{d^2U}{dz^2} + \frac{k}{\omega}\left[\frac{D'}{D} - \beta - \frac{F'}{F}\right]\frac{dU}{dz} + \left(\frac{\beta g}{\omega^2} - E\right)\frac{F}{D}\right] u_z = 0 , \tag{2.23}$$

where $(D', F') \equiv \frac{d}{dz}(D, F)$, $A^2 = \frac{B^2}{\mu\varrho_0}$ is the square of the Alfvén velocity, $\omega = \sigma - kv$ is the Doppler shifted frequency and

$$D(z) \equiv \left(1 - \frac{k^2A^2}{\omega^2}\right) \tag{2.24}$$

$$E(z) \equiv 1 - \frac{A^2}{\omega^2}\left(k^2 - \frac{\sigma^2}{c^2}\right) \tag{2.25}$$

$$F(z) \equiv \frac{(k^2 + l^2) - \frac{k^2A^2}{\omega^2}\left(k^2 + l^2 - \frac{\sigma^2}{c^2}\right)}{1 - \frac{A^2}{\omega^2}\left(k^2 - \frac{\sigma^2}{c^2}\right)} . \tag{2.26}$$

We next assume $u_z = \varphi(z) \exp\left(\frac{\beta}{2} z\right)$ with $\beta = -\varrho_0'/\varrho_0 > 0$. Consequently, (2.23) becomes

$$G\varphi''(z) + G'\varphi'(z) + \left\{\frac{Gk}{\omega}\frac{d^2U}{dz^2} + \left(\frac{\beta}{2} + \frac{k}{\omega}\frac{dU}{dz}\right)(G' - \beta G) + \frac{G\beta^2}{4} + \\ + \left[\frac{N^2}{\omega^2} - 1 + \frac{A^2}{\omega^2}\left(k^2 - \frac{\sigma^2}{c^2}\right)\right]\right\}\varphi = 0 , \tag{2.27}$$

where

$$\left.\begin{aligned} G \equiv G(z) = \frac{D}{F} &= \frac{(\omega^2 - \omega_a^2)(\omega^2 - \omega_a^2 + M^2\sigma^2)}{\omega^2[(k^2 + l^2)(\omega^2 - \omega_a^2) + M^2\sigma^2k]} , \\ \omega_a^2 = k^2A^2 , &\qquad M^2 = A^2/c^2 . \end{aligned}\right\} \tag{2.28}$$

When the basic flow is uniform, the coefficients of (2.27) are constant and the equation has solution φ which is proportional to $\exp(imz)$ where m is a constant vertical wavenumber. However, when the basic flow is not uniform, the most striking feature of (2.27) is its singularities at

$$\omega = 0, \pm\omega_a, \pm\lambda, \pm\delta , \tag{2.29}$$

where

$$\left.\begin{aligned} \omega_a^2 &\equiv k^2A^2 , \qquad \lambda \equiv \omega_a\left[1 - \frac{\sigma^2}{c^2(k^2 + l^2)}\right]^{1/2} , \\ \delta &\equiv (\omega_a^2 - M^2\sigma^2)^{1/2} . \end{aligned}\right\} \tag{2.30}$$

These singularities may evidently be regarded as a consequence of the loss of higher-order derivatives due to the neglect of dissipative effects. In order to eliminate these singularities, it is necessary to include the dissipating effects such as kinematic viscosity, magnetic viscosity and heat conduction. Physically, these singular points correspond to critical (singular) levels which are nothing but resonance levels where there will be maximum absorption of energy.

Thus the governing wave equation (2.27) is singular at heights where the flow velocity U assumes special values such that the Doppler shifted frequency $\omega = 0$, $\pm\omega_a$, $\pm\lambda$ and $\pm\delta$ in contrast to the singularities at $\omega = 0$, $\pm\omega_a$ for the case of no displacement current. Evidently, the effect of the displacement current is to increase the number of critical layers.

3. Solutions of the wave equation

We follow BOOKER and BRETHERTON's (1967) analysis to resolve the singularities of the wave equation (2.27). For this the power-series solutions of (2.27) near the critical levels are obtained by using the method of Frobenius and assuming that the shear velocity dU/dz is independent of height. Near the critical level $\omega = 0$, the complete solution of (2.27) is of the form

$$\varphi = [A_c + B_c \log (z - z_c)]\,[1 + c_1(z - z_c) + \ldots] + \\ + B_c \sum_{r=0}^{\infty} d_r(z - z_c)^{-r} , \tag{3.1}$$

where A_c and B_c are constants of integration, d_r (or c_0) and $c_1, c_2, c_3, \ldots$ are known constants. Similar solutions also exist near other critical levels corresponding to singularities stated in (2.29). The wave has a discontinuity at the above critical level.

To determine the gross attenuation of a wave when it passes through all of the above seven critica levels, we obtain the solution of the wave equation (2.27) away from all these critical levels and calculate the gross attenuation. For this we restrict the value of ω by an approximation

$$\omega^2 \ll N^2 \tag{3.2}$$

where $N = (g\beta)^{1/2}$ is the Brunt-Väisälä frequency.

Since solutions away from the critical levels are of special interest, we also assume $\omega^2 > \omega_a^2$ and hence absorption takes place only at the five critical levels since $\omega^2 > \omega_a^2$.

In view of the above approximation and the transformation $(z - z_c)^{-1} = \zeta$, the wave equation (2.27) takes the form

$$\zeta^4 \frac{d^2\varphi}{d\zeta^2} + 2\zeta^3 \frac{d\varphi}{d\zeta} + (N^2 - M^2\sigma^2)(k^2 + l^2)\, q^{-2}\zeta^2\varphi = 0 , \tag{3.3}$$

where $q^2 = \left(k \frac{dU}{dz}\right)^2$.

The solution of (3.3) around $\zeta = 0$ is

$$\varphi(z) = A_1(z - z_c)^{1/2 + i\mu}\,[1 + a_{11}(z - z_c)^{-1} + \ldots] + \\ + B_1(z - z_c)^{1/2 - i\mu}\,[1 + b_{11}(z - z_c)^{-1} + \ldots] , \tag{3.4}$$

where A_1, B_1 are constants of integration and $a_{11}, b_{11}, \ldots$ are known constants; and

$$\mu = (J - \tfrac{1}{4})^{1/2} , \qquad J = (N^2 - M^2\sigma^2)(k^2 + l^2)\, q^{-2} . \tag{3.5ab}$$

If we fix the branch of the complex power in (3.4) by setting

$$\varphi(z) = \\ = \left\{\begin{aligned} &A_1|z - z_c|^{1/2} \exp[i\mu \log|z - z_c|]\,[1 + a_{11}(z - z_c)^{-1} + \ldots] + \\ &+ B_1|z - z_c|^{1/2} \exp[-i\mu \log|z - z_c|] \times \\ &\times [1 + b_{11}(z - z_c)^{-1} + \ldots] \quad \text{for} \quad z > z_c \\ &iA_1|z - z_c|^{1/2} \exp[\pi\mu + i\mu \log|z - z_c|] \times \\ &\times [1 + a_{11}(z - z_c)^{-1} + \ldots] + \\ &+ iB_1|z - z_c|^{1/2} \exp[-\pi\mu - i\mu \log|z - z_c|] \times \\ &\times [1 + b_{11}(z - z_c)^{-1} + \ldots] \quad \text{for} \quad z < z_c . \end{aligned}\right\} \tag{3.6ab}$$

It is noted that the amplitude above and below the point $z = z_c$ differ by a factor of $\exp(\pi\mu)$.

4. Critical layer absorption of waves

We study the absorption of waves near the critical levels using momentum transfer to the mean flow and the group velocity approach.

(i) *Momentum transfer to the mean flow*

Solutions of the wave equation may be interpreted locally as upward and downward travelling waves from an energy consideration. The upward transfer of the wave energy per unit area at any level is equal to the mean rate of working of pressure forces on the fluid, $\tilde{E} = \overline{pu_z}$ where the bar denotes an average over a horizontal wave length. But

$$\tilde{E} = \overline{pu_z} = \tfrac{1}{4}(pu_z^* + p^*u_z) = \tfrac{1}{2}\,\mathrm{Re}\,(pu_z^*) . \tag{4.1}$$

Thus

$$\tilde{E} = \frac{\varrho_c \omega}{2F} \operatorname{Re}(i\psi\psi^*), \tag{4.2}$$

where the quantity with an asterick denotes the complex conjugate.

Application of the transformation

$$\psi(z) = \varphi(z) \exp\left[\int_0^z (G'/2G)\, dz\right], \tag{4.3}$$

reduces (4.2) into the form

$$\tilde{E} = \frac{\varrho_c \omega}{2F} \operatorname{Re}\left[i\varphi'\varphi^* \exp\left\{\int_0^z (G'/2G)\, dz\right\}\right]. \tag{4.4}$$

Thus the expression for $\tilde{E}$ in terms of A_1 and B_1 can be obtained using (3.6) as

$$\tilde{E} = \begin{cases} -\dfrac{\varrho_0}{2F} e^{\beta z} \mu\omega[|A_1|^2 - |B_1|^2] & \text{for} \quad z > z_c, \\ -\dfrac{\varrho_0}{2F} e^{\beta z} \mu\omega[|B_1|^2 \exp(-2\pi\mu) - |A_1|^2 \exp(2\mu\pi)] & \text{for} \quad z < z_c. \end{cases} \tag{4.5ab}$$

In view of (4.5ab), for the first term in (3.1), the energy flux, $\tilde{E}$ is positive both above and below the critical levels. Hence the energy flows upward with time and the associated wave is an upward travelling wave. While for the second term in (3.1), $\tilde{E}$ is negative and hence it corresponds to downward travelling wave.

We next define the wave action flux $\tilde{M} = \tilde{E}/\omega$ as the ratio of the energy flux to the level relative frequency, which is proportional to the rate of transfer of winddirected momentum up the velocity gradient. We have $\tilde{M} = \frac{\varrho_c}{2F} \operatorname{Re}[i\psi'\psi^*]$ and it can be shown that $\frac{d\tilde{M}}{dz} = 0$. Hence $\tilde{M}$ is conserved energywhere in the fluid except at the critical levels and is taken as a convenient measure of the intensity of the wave.

$$\frac{\tilde{M}_{\text{incident}}}{\tilde{M}_{\text{transmitted}}} = \exp\left[2\pi\left\{\frac{(N^2 - M^2\sigma^2)\,(k^2 + l^2)}{q^2} - \frac{1}{4}\right\}^{1/2}\right]. \tag{4.6}$$

(ii) *Group velocity near the critical levels*

In the previous section the phenomenon of absorption of waves at the critical levels has been discussed by using the concept of momentum transport. This phenomenon can also be explained by using the concept of group velocity. We consider a slowly varying medium with k and l as horizontal wavenumbers and m as the vertical wavenumber.

The wave equation (2.27) can be rewritten in the form

$$\psi'' + \left[\frac{\beta}{2}\left(\frac{G'}{G} - \beta\right) + \frac{\beta^2}{4} + \frac{G'^2}{4G^2} + \right.$$
$$\left. + \frac{1}{G}\left\{\frac{N^2}{\omega^2} - 1 + \frac{A^2}{\omega^2}\left(k^2 - \frac{\sigma^2}{c^2}\right)\right\}\right]\psi = 0, \tag{4.7}$$

where $\psi(z) = \varphi(z) \int_0^z \left(\frac{G'}{2G}\right) dz$.

We consider a wave packet as a localized disturbance with reasonably well defined dominant frequency and wavenumbers. This wave packet moves with the group velocity defined by

$$C_g = \left(\frac{\partial\sigma}{\partial k}, \frac{\partial\sigma}{\partial l}, \frac{\partial\sigma}{\partial m}\right). \tag{4.8}$$

The wave with the wavenumbers (k, l, m) satisfies the dispersion relation

$$-\left(m^2 + \frac{\beta^2}{4}\right)(\omega^2 - k^2A^2)\left[(\omega^2 - A^2\left(k^2 - \frac{\sigma^2}{c^2}\right)\right] +$$
$$+ \left[\omega^2(k^2 + l^2) - k^2A^2\left(k^2 + l^2 - \frac{\sigma^2}{c^2}\right)\right] \times$$
$$\times \left[N^2 - \omega^2 + A^2\left(k^2 - \frac{\sigma^2}{c^2}\right)\right] = 0. \tag{4.9}$$

This equation can be solved for m^2 to obtain

$$m^2 = \frac{[(k^2 + l^2)(\omega^2 - \omega_a^2) + M^2\sigma^2k^2][N^2 - \omega^2E]}{(\omega^2 - \omega_a^2)\,\omega^2E} - \frac{\beta^2}{4}, \tag{4.10}$$

where

$$\omega^2E = (\omega^2 - \omega_a^2 + M^2\sigma^2). \tag{4.11}$$

Evidently, as the wave group approaches the critical level $\omega^2 \to \omega_a^2$, the vertical wavenumber m increases indefinitely and is given asymptotically by

$$m^2(\omega^2 - \omega_a^2) \approx k^2(N^2 - M^2\sigma^2). \tag{4.12}$$

The vertical component, $\partial\sigma/\partial m$ of the group velocity at any level is given by

$$\frac{\partial\sigma}{\partial m} = \frac{2m(\omega^2 - \omega_a^2)\,\omega^2E}{-(2\omega + 2\sigma M^2)\,D_1 - \omega^2E[2\omega(k^2 + l^2 + m^2 + \beta^2/4) + 2\sigma k^2M^2] + N^2[k^2(2\omega + 2\sigma M^2) + 2\omega l^2]}, \tag{4.13}$$

where

$$D_1 = [(\omega^2 - \omega_a^2)(k^2 + l^2 + m^2 + \beta^2/4) + M^2k^2\sigma^2]. \tag{4.14}$$

The asymptotic behaviour of $\partial\sigma/\partial m$ corresponding to (4.12) is given by

$$\frac{\partial\sigma}{\partial m} \sim -\frac{k^2(N^2 - M^2\sigma^2)}{\omega}(z - z_c)^{3/2}. \tag{4.15}$$

Thus when $|z - z_c|$ is small the height of the wave packet satisfies the equation

$$\frac{dz}{dt} = a(z - z_c)^{3/2}, \tag{4.16}$$

where a is a constant. Integration gives

$$(z - z_c)^{1/2} = -\frac{2}{at + b}, \tag{4.17}$$

where b is another integrating constant. The wave packet thus slows down in such a way that it never reaches the critical level $z = z_c$ in a finite time. It is therefore neither transmitted nor reflected.

The other two components of the group velocity are given by

$$\frac{\partial\sigma}{\partial k} = \frac{-[2\omega U + 2kA^2]\left[D_1 + \omega^2E\left(k^2 + l^2 + m^2 + \frac{\beta^2}{4}\right) - N^2(k^2 + l^2)\right] - 2kE\omega^2(N^2 - E\omega^2)}{-(2\omega + 2\sigma M^2)\,D_1 - \omega^2E\left[2\omega\left(k^2 + l^2 + m^2 + \frac{\beta^2}{4}\right) + 2\sigma M^2k^2\right] + N^2[k^2(2\omega + 2\sigma M^2) + 2\omega l^2]} \tag{4.18}$$

and

$$\frac{\partial\sigma}{\partial l} = \frac{-2l(\omega^2 - \omega_a^2)(N^2 - \omega^2E)}{-(2\omega + 2\sigma M^2)\,D_1 - \omega^2E\left[2\omega\left(k^2 + l^2 + m^2 + \frac{\beta^2}{4}\right) + 2\sigma M^2k^2\right] + N^2[k^2(2\omega + 2\sigma M^2) + 2\omega l^2]}. \tag{4.19}$$

As the critical level is approached,

$$\frac{\partial\sigma}{\partial k} \to U + A\,, \quad \frac{\partial\sigma}{\partial l} \to 0\,. \tag{4.20ab}$$

Therefore the wave group is effectively captured in this neighborhood and constrained thereafter to propagate along the mean flow.

5. Discussions and conclusions

The propagation of Alfvén waves has been studied in an isothermal atmosphere when the displacement current is included in this analysis. It is shown that the governing wave equation is singular at heights where the flow velocity U assumes special values such that the Doppler shifted frequency $\omega = 0$, $\pm\omega_a$, $\pm\lambda$ and $\pm\delta$. When the displacement current is neglected, the singularities correspond to $\omega = 0$ and $\pm\omega_a$. Thus the effect of displacement current is to increase the number of singularities of the equation, and, in turn, is to reduce the gross attenuation of the wave across the critical levels.

In the limit $c \to \infty$, the critical values $\pm\lambda$ and $\pm\delta$ coincide with the critical values $\pm\omega_a$.

Since the wave action flux, $\tilde{M}$ is constant everywhere except at the critical levels, we found it as an appropriate measure of the magnitude of the waves. Using this wave action flux as a measure of intensity of the wave, the attenuation of waves at the critical levels is examined.

It is shown that the attenuation that a wave suffers on passage through the critical levels $\zeta = 0$ (that is, away from all the above seven critical levels) in an isothermal medium is less than the attenuation that a wave suffers on passage through the critical levels in a fluid in which the displacement current is neglected. This result is confirmed using the group velocity approach. Finally, it is confirmed that the effect of the displacement current is to increase the number of critical layers and in turn, is to reduce the gross attenuation of waves across the critical levels.

Acknowledgement

Authors express their grateful thanks to Prof. LANCKAU for suggesting some corrections and improvements of the paper.

References

1 BOOKER, J. R., BRETHERTON, F. P.: The critical layer for internal gravity waves in a shear flow. J. Fluid Mech. 27 (1967), 513–539.
2 RUDRAIAH, N.; VENKATACHALAPPA, M.: Propagation of internal gravity waves in perfectly conducting fluids with shear flow, rotation and transverse magnetic field. J. Fluid Mech. 52 (1972), 193–206.
3 ACHESON, D. J.: The critical level for hydromagnetic waves in a rotating fluid. J. Fluid Mech. 53 (1972), 401–415.
4 MCKENZIE, J. F.: On the existence of critical levels with applications to hydromagnetic waves. J. Fluid Mech. 58 (1973), 709–726.
5 ELTAYEB, I. A.; KANDASWAMY, P.: Critical levels in a vertically rotating compressible atmosphere. Quart. J. Mech. and Appl. Math. 32 (1979), 313 to 330.

Received September 28, 1987, revised version March 24, 1988

Addresses: Dr. V. SRINIVASAN, Dr. P. KANDASWAMY, Department of Mathematics, Bharathiar University, Coimbatore 641046, Tamil Nadu, India; Prof. Dr. LOKENATH DEBNATH, Department of Mathematics, University of Central Florida, Orlando, FL 32816, U.S.A.

ZAMM Z. angew. Math. Mech. 69 (1989) 2, 100–104

SILVA-FREIRE, A. P.

On the Matching Conditions for a Two-Deck Compressible Turbulent Boundary Layer Model

I. Introduction

Over the past forty years the problem of compressible turbulent boundary layers has been the subject of many theoretical and experimental studies. The difficulty of treating both turbulence and compressibility, and the success in describing incompressible turbulent boundary layers, has motivated a number of authors to look for features of the incompressible flow which are preserved in compressible flow. These identified common features, together with the procedure for incompressible flow, and adequate modifications, can then be used to try to develop a theory for compressible flow.

Recently, some authors (MELLOR [12], YAJNIK [15] and others) have developed formal asymptotic theories for incompressible turbulent boundary layers, employing the method of matched asymptotic expansions together with asymptotic hypotheses describing the order of the various terms in the equations of mean motion. The theories consider the entire fluid region and avoid similarity, dimensional or eddy viscosity arguments. The main idea underlying the theories is that there are two different layers whose properties and thickness can be described by their respectives length scales.

Using the same idea, AFZAL [2] attempted to extend the asymptotic theories developed for incompressible flow to compressible flow. He formulated a higher order theory for compressible turbulent boundary layer flow of a perfect gas with constant specific heats when $(\gamma - 1)\, M_\infty^2$ and molecular Prandtl number are of order unity; he worked with an underdetermined system of equations of mean motion and showed that the structure of the solution is the same as for incompressible flow, i.e., two layers are required. However, several authors, MELNIK and GROSSMAN [13], ADAMSON and FEO [1], LIOU and ADAMSON [10], pointed out that although AFZAL had presented the solution by limit-function expansions, he had worked with an underdetermined system of equations and had not showed that the wall and the defect solutions did, in fact, match. Thus, according to their analysis the fact that the density varied by order unity from wall to free stream values across the boundary layer caused difficulties which could not be solved by AFZAL's solution. It is of interest to note that the derivation of this result strongly depends on AFZAL's assumed form of the asymptotic expansions, and that no systematic study about the domains of validity of the solutions was made. Thus, those analyses cannot be seen as conclusive proof that the defect and wall layers do not match.

The aim of this work is to carry out a study of the domains of validity of the defect and wall layer solutions using the intermediate variable technique. This technique was introduced by KAPLUN in the mid fifties and reviewed by LAGERSTROM and CASTEN in the early seventies. The intermediate variable technique provides a more rigorous mathematical treatment to deal with problems where perturbation techniques can be applied. In this theory, heuristic ideas are used to develop a systematic method of obtaining domains of validity of asymptotic expansions by a study of the corresponding equations. The theory emphasizes the idea of characterizing expansions by their domains of validity, providing not only a more basic understanding for the matching but also a deeper insight into the construction of the expansions. This is illustrated in the work of LAGERSTROM and CASTEN [8], which also shows how it can be applied to different problems.

II. Equations of mean motion and asymptotic hypotheses

In this section we introduce the two-dimensional Navier-Stokes equations of mean motion for a compressible fluid using the mass-weighted-averaging procedure and follow AFZAL's analysis of KISTLER's data to estimate the order of magnitude of the turbulent fluctuations of velocity, pressure, temperature and density.

The continuity, momentum, energy and state equations obtained by mass-weighted averaging can be written in the following non-dimensional form:

a) Continuity

$$\frac{\partial}{\partial x_j}(\bar{\varrho}\tilde{u}_j) = 0 \tag{1}$$

b) Momentum

$$\frac{\partial}{\partial x_j}(\bar{\varrho}\tilde{u}_i\tilde{u}_j) = -\frac{\partial\bar{p}}{\partial x_i} + \frac{\partial}{\partial x_j}(-\overline{\varrho u_i'u_j'} + R^{-1}\overline{\tau_{ij}}) \tag{2}$$

c) Energy

$$\frac{\partial}{\partial x_j}(\bar{\varrho}\tilde{t}\tilde{u}_j) = D\tilde{u}_j \frac{\partial \bar{p}}{\partial x_j} + D\overline{u'_j \frac{\partial p}{\partial x_j}} + \frac{\partial}{\partial x_j}(R^{-1}\sigma^{-1}\overline{q_j} - \overline{\varrho t' u'_j}) + R^{-1}D\overline{\tau_{ij}\frac{\partial u_i}{\partial x_j}}, \quad (3)$$

d) State

$$\bar{p} = \frac{\gamma - 1}{\gamma}\frac{1}{D}\bar{\varrho}\tilde{t}, \quad (4)$$

where the stress tensor τ_{ij} and the heat flux tensor q are given respectively by

$$\tau_{ij} = \lambda\delta_{ij}\frac{\partial u_l}{\partial x_l} + \mu\left(\frac{\partial u_i}{\partial x_j} + \frac{\partial u_j}{\partial x_i}\right), \quad (5)$$

and

$$q_j = -K\frac{\partial t}{\partial x_j}. \quad (6)$$

In these equations, x, u, p, ϱ, and t have their classical meaning, λ is the bulk viscosity ($= -2/3\mu$), μ the dynamic viscosity, K the thermal viscosity, and δ_{ij} the Kronecker delta, having the value 1 for $i = j$ and 0 for $i \neq j$. All lengths are non-dimensionalized by a typical body dimension L, velocities by the characteristic reference velocity U_∞, pressure by $\varrho_\infty U_\infty^2$, temperature by T_∞, density by ϱ_∞ and viscosity by μ_∞. The quantity $R = U_\infty\varrho_\infty L/\mu_\infty$ is the characteristic Reynolds number, M_∞ is a characteristic Mach number, $D = (\gamma - 1) M_\infty^2$ is the compressibility factor, σ is the molecular Prandtl number and γ is the ratio of specific heats. The superscript denotes turbulent fluctuations, the bar denotes conventional time averaging and the tilde denotes mass-weighted averaging. A summation is understood for repeated indices. For further details concerning these equations the reader is referred to either CEBECI and SMITH [4] or LIEPMANN and ROSHKO [9].

Based on the measurements of KISTLER [6], AFZAL concluded that the fluctuation u' can be assumed to be of the order of the friction velocity divided by the mean free stream velocity, i.e.,

$$u' = O\left(\frac{u_\tau}{U_\infty}\right) = O\left(\frac{1}{U_\infty}\sqrt{\frac{\tau_w}{\varrho_w}}\right) = O(\varepsilon). \quad (7)$$

For the static pressure, density and temperature fluctuations, further measurements of KISTLER and CHEN [7] and of MORKOVIN [14] have shown that (a) u', ϱ' and t' have the same order of magnitude and (b) the root-mean-squate value of p' is proportional to u'.

The notation for turbulent Reynolds terms employed here will be the same as presented by AFZAL, i.e., second order correlations of the type $\overline{a'b'}$ are denoted by τ_{ab} and third-order correlations of the type $\overline{a'b'c'}$ by τ_{abc}. Thus a correlation of the type $\overline{-b'\partial a'/\partial y}$ is denoted by $\overline{-a'_y b'}$. In view of the above remarks, the scales of fluctuations can be written without loss of generality as

$$O(u') = O(v') = O(\varrho') = O(t') = O(\varepsilon) \quad (8)$$

$$O(p') = \varepsilon^2. \quad (9)$$

III. Analysis of the domains of validity of the asymptotic expansions

In this section we apply the intermediate variable technique to the problem of compressible turbulent boundary layers. This requires the knowledge of the concepts of domain of validity of asymptotic expansions, overlap, formal validity of equations, limit process, etc. Most of these concepts lie in ideas of the late SAUL KAPLUN and his co-workers, and have been presented with particular clarity in an article of LAGERSTROM and CASTEN [8]. Basic concepts such as those of asymptotic expansion, order classes, uniform convergence on a function class and uniform validity on a function class will not be introduced here since they have been extensively treated in the literature. The order to which a function $\eta(\varepsilon)$ belongs will be denoted by ord η.

Before we proceed, let us remember that the parameters σ and D are assumed to be order unity, and that the boundary layer has three length scales [AFZAL, 2]: an inviscid scale L, an outer scale $\delta = \varepsilon L$, and an inner scale $\hat{\delta} = \nu/u_\tau = \hat{\varepsilon}\delta$. The ratio of the outer to the inviscid length scale is $\delta/L = \varepsilon$ and of the inner to the outer scale is

$$\hat{\delta}/\delta = \nu/\delta u_\tau = \hat{\varepsilon}. \quad (10)$$

Moreover, parameters ε and $\hat{\varepsilon}$ are related by the well known turbulent skin friction equation which yields [MELLOR, 12]

$$\varepsilon = O((\ln\hat{\varepsilon})^{-1}), \quad \text{as} \quad \varepsilon \to 0. \quad (11)$$

Our objective is to obtain an approximation for the system of equations (1—6) which is uniformly valid in a given domain. For this reason, it is necessary to define the concept of uniform domain of validity for such approximation. Here, our first difficulty occurs, since, although we can formalize clearly the concept of uniform validity on a function class, i.e., when solutions are close, the basic difficulty of deciding when equations are close still remains unanswered.

Therefore in order to introduce KAPLUN's matching principle it is firstly necessary to define η-limit and the domain of validity of equations. The η-limit of an equation $E(x, y; \varepsilon)$ is defined as follows. Let the intermediate variable y_η be

$$\eta(\varepsilon)\, y_\eta = y, \quad (12)$$

where, as indicated, η is a function of ε. Then, the η-limit of $E(x, y; \varepsilon)$ is

$$\lim_\eta E(x, y; \varepsilon) = \lim E(x, \eta(\varepsilon)\, y_\eta; \varepsilon) \quad \text{as} \quad \varepsilon \to 0 \quad (13)$$

with y_η fixed.

Note that $\eta(\varepsilon)$, as defined by equation (12), is in fact a stretching function. If we substitute (12) into the motion equation and arbitrarily vary the order of magnitude of $\eta(\varepsilon)$, we can then use the definition of η-lim to study the effect on the motion equations. This will reveal the importance of the various terms of the equation of motion in the different flow regions. Now, if we introduce (12) into the system (1—6), each term will have a formal order in ε depending on the assumptions in section II and on expressions (10) and (11). The derivatives and y_η are formally considered to be of order unity. For example, we note that the terms $(\varrho\tau_{uv})_y$ and μu_{yy}, in equation (2), are formally of order ε^2/η and $\varepsilon^2\hat{\varepsilon}/\eta$ respectively. Passing the η-limit in the equations resulting from (2—3), we find the following formal limits depending on different order of magnitude of η:

x-momentum equation:

ord $\eta > 1$: $\quad (\bar{\varrho}\tilde{u})\,\tilde{u}_x + (\bar{\varrho}\tilde{v})\,\tilde{u}_y + \bar{p}_x = 0, \quad$ (14a)

ord $\eta = 1$: $\quad (\bar{\varrho}\tilde{u})\,\tilde{u}_x + (\bar{\varrho}\tilde{v})\,\tilde{u}_y + \bar{p}_x = 0, \quad$ (14b)

ord $\varepsilon <$ ord $\eta < 1$: $\quad (\bar{\varrho}\tilde{u})\,\tilde{u}_x + (\bar{\varrho}\tilde{v})\,\tilde{u}_y + \bar{p}_x = 0, \quad$ (14c)

ord $\eta =$ ord ε^2: $\quad (\bar{\varrho}\tilde{u})\,\tilde{u}_x + (\bar{\varrho}\tilde{v})\,\tilde{u}_y + \bar{p}_x = (\bar{\varrho}\tau_{uv})_y, \quad$ (14d)

ord $\varepsilon^3 <$ ord $\eta <$ ord ε^2: $\quad (\bar{\varrho}\tau_{uv})_y = 0, \quad$ (14e)

ord $\hat{\varepsilon} <$ ord $\eta <$ ord ε^3: $\quad (\bar{\varrho}\bar{\tau}_{uv})_y = 0, \quad$ (14f)

ord $\eta =$ ord $\hat{\varepsilon}$: $\quad (\bar{\mu}\tilde{u}_y)_y + (\bar{\varrho}\bar{\tau}_{uv})_y = 0, \quad$ (14g)

ord $\eta <$ ord $\hat{\varepsilon}$: $\quad (\bar{\mu}\tilde{u}_y)_y = 0; \quad$ (14h)

energy equation:

ord $\eta > 1$: $\quad \bar{\varrho}\tilde{u}\tilde{t}_x + \bar{\varrho}\tilde{v}\tilde{t}_y - D(\tilde{u}\bar{p}_x + \tilde{v}\bar{p}_y) = 0, \quad$ (15a)

ord $\eta = 1$: $\quad \tilde{\varrho}\tilde{u}\tilde{t}_x + \bar{\varrho}\tilde{v}\tilde{t}_y - D(\tilde{u}\bar{p}_x + \tilde{v}\bar{p}_y) = 0, \quad$ (15b)

ord $\varepsilon^2 <$ ord $\eta < 1$: $\quad \bar{\varrho}\tilde{u}\tilde{t}_x + \bar{\varrho}\tilde{v}\tilde{t}_y - D(\tilde{u}\bar{p}_x + \tilde{v}\bar{p}_y) = 0, \quad$ (15c)

ord $\eta =$ ord ε^2: $\quad \bar{\varrho}\tilde{u}\tilde{t}_x + \bar{\varrho}\tilde{v}\tilde{t}_y - D(\tilde{u}\bar{p}_x + \tilde{v}\bar{p}_y) = (\bar{\varrho}\bar{\tau}_{vt})_y + D(\bar{\tau}_{vp})_y, \quad$ (15d)

ord $\varepsilon^3 <$ ord $\eta <$ ord ε^2: $\quad (\bar{\varrho}\bar{\tau}_{vt})_y + D(\bar{\tau}_{vp})_y = 0, \quad$ (15e)

ord $\hat{\varepsilon} <$ ord $\eta <$ ord ε^3: $\quad (\bar{\varrho}\bar{\tau}_{vt})_y + D(\bar{\tau}_{vp})_y = 0, \quad$ (15f)

ord $\eta =$ ord $\hat{\varepsilon}$: $\quad \sigma^{-1}(\bar{\mu}\tilde{t}_y)_y + (\bar{\varrho}\bar{\tau}_{vt})_y + D(\bar{\tau}_{vp})_y = 0, \quad$ (15g)

ord $\eta <$ ord $\hat{\varepsilon}$: $\quad \sigma^{-1}(\bar{\mu}\tilde{t}_y)_y + 0, \quad$ (15h)

The two sets of equations, (14 a—h) and (15 a—h), are the two important ones. Indeed, the continuity equation, y-momentum equation and the state equation result in trivial sets of equations. For instance, the y-momentum equation, whatever is the order of η, results in

$$\bar{p}_y = 0. \quad (16)$$

We also note that if $\lim_\eta$ applied to equations (2—3) yields (14e—15e), then the formal η-limit of (14d—15d) also yields

(14e—15e). Thus according to LAGERSTROM and CASTEN, equations (14d—15d) are "rich enough" to contain (14e—15e). The two definitions below are due to LAGERSTROM and CASTEN:

Definition I. If E is an equation and $\lim_{\eta_1} E = E_1$, $\lim_{\eta_2} E = E_2$, and also $\lim_{\eta_2} E_1 = E_2$, we say that E_1 contains E_2 (relative to E).

Definition II. The formal domain of validity of an equation F, relative to the full equation E, is ord η such that $\lim_\eta E$ is either F or an equation contained in F.

These definitions imply that the formal domain of validity of (14d) and (15d) is given by

$$D_0 = \{\eta/\text{ord}\,\eta > \text{ord}\,\hat{\varepsilon}\}\,, \tag{17}$$

and that of (14g) and (15g) is given by

$$D_i = \{\eta/\text{ord}\,\eta < \text{ord}\,\varepsilon^2\}\,. \tag{18}$$

Equations d's and g's are called the principal equations. They are, in fact, the important equations. One can observe that equations d's contain equations a's, b's, c's, e's and f's; equations g's contain equations e's, f's and h's but neither of them is contained in the other.

Principle. If f is a solution of an equation E, and E^* is an approximate equation, then there exists a solution f^* of E^* whose actual domain of validity (as an approximation to f) includes the formal domain of validity of E^* (as an approximation to E).

In the LAGERSTROM and CASTEN article, important modifications of this principle are discussed since it ceases to be valid when small terms have large integrated effects. Fortunately, terms such as $\varepsilon \ln \varepsilon$, which appeared in the solution of LAGERSTROM and CASTEN's second mode equation, are not expected to occur here. These terms provoke an effect of larger magnitude than the order formally indicated which cannot be compensated by the lower order terms leading to the phenomenon of integrated effects and resulting switchback. So, according to the above remarks, it seems that KAPLUN's principle can be applied to the problem defined by equations (1—6). These results yields, together with results (17—18), that the domain of overlap is given by

$$D = D_0 \cap D_i = \{\eta/\text{ord}\,\hat{\varepsilon} < \text{ord}\,\eta < \text{ord}\,\varepsilon^2\}\,. \tag{19}$$

Equations (14 a—h) illustrate the well known balance among inertia, pressure, Reynolds stress and viscous stress in the flow. In fact, equation (14h) shows that very close the wall the effects due to the viscous shear stress are the dominating effects; as the distance from the wall increases, equation (14g) shows that the Reynolds shear stress becomes larger and balances the viscous shear stress, both effects being of lower order than the pressure gradient effects. Further away, the overlap region is reached, and equations (14 e—f) show that the Reynolds stress effects are the dominating effects. Above this region, equation (14d) shows that the inertia, pressure and Reynolds stress are of equal importance. Finally, in the outermost layer, comprising most of the boundary layer, equations (14 a—c) show that the inertia and pressure effects are the dominating effects.

Of course, a similar type of analysis applied to the energy equation emphasizes, near the wall, the conduction terms and, further away from the wall, the turbulent terms.

IV. Asymptotic analysis

In this section, we apply the matched asymptotic expansion method to the problem of compressible, turbulent, boundary layers to show how solutions for the defect and wall layers can be obtained which do match in an overlap domain. Since most of the analysis concerning the obtaining of the asymptotic expansion is conventional, only a few comments about it will be made here. The bars and tildes will be omitted in all subsequent expressions.

Defect layer

For this region we write

$$u(x, y; R) = u_1(x, y) + \varepsilon u_2(x, y) + \varepsilon^2 u_3(x, y) + \dots\,, \tag{20a}$$

$$v = \varepsilon v_1(x, y) + \varepsilon^2 v_2(x, y) + \varepsilon^3 v_3(x, y) + \dots\,, \tag{20b}$$

$$p = p_1(x, y) + \varepsilon p_2(x, y) + \varepsilon^2 p_3(x, y) + \dots\,, \tag{20c}$$

$$\varrho = \varrho_1(x, y) + \varepsilon \varrho_2(x, y) + \varepsilon^2 \varrho_3(x, y) + \dots\,, \tag{20d}$$

$$t = t_1(x, y) + \varepsilon t_2(x, y) + \varepsilon^2 t_3(x, y) + \dots\,, \tag{20e}$$

$$\tau_{ab} = \varepsilon^2 \tau_{ab_1}(x, y) + \varepsilon^3 \tau_{ab_2}(x, y) + \varepsilon^4 \tau_{ab_3}(x, y) + \dots\,, \tag{20f}$$

$$\tau_{abc} = \varepsilon^3 \tau_{abc_1}(x, y) + \varepsilon^4 \tau_{abc_2}(x, y) + \varepsilon^5 \tau_{abc_3}(x, y) + \dots\,. \tag{20g}$$

where $y = Y/\delta(R)$ is the stretched defect layer variable.

Introduction of these expressions into the equations of mean motion and collection of coefficients of various powers of ε give the equations for the successive approximations. These equations are, except for a few differences arising from distinct averaging procedures, the same as those obtained by AFZAL. They show that the pressure is constant across the outer layer to order ε and that viscous and heat-conduction terms do not occur in the outer layer. From the matching between the inviscid and outer expansions (see AFZAL), it follows that

$$\left.\begin{aligned} &u_1(x, y) = U_1(x, 0)\,, && p_1(x, y) = P_1(x, 0)\,, \\ &\varrho_1(x, y) = \mathrm{P}_1(x, 0)\,, && t_1(x, y) = T_1(x, 0)\,, \qquad y \to \infty\,, \\ &\tau_{ab_1}(x, y) = \mathrm{T}_{ab_1}(x, 0)\,, && \tau_{abc_1}(x, y) = \mathrm{T}_{abc_1}(x, y)\,. \end{aligned}\right\} \tag{21}$$

Here, capital letters are used to denote terms valid in the inviscid region and, in order to follow AFZAL's notation, the quantities $U_1(x, 0)$, $P_1(x, 0)$, $\mathrm{P}_1(x, 0)$ and $T_1(x, 0)$ are denoted by U_{10}, P_{10}, P_{10} and T_{10} respectively. The solution of the lowest order equation which satisfies the matching conditions is

$$\left.\begin{aligned} &u_1 = U_{10}\,, \quad \varrho_1 = \mathrm{P}_{10}\,, \\ &v_1 = [Q - (\mathrm{P}_{10} U_{10x})]/\mathrm{P}_{10}\,, \\ &p_1 = P_{10}\,, \quad t_1 = T_{10}\,. \end{aligned}\right\} \tag{22}$$

where Q is a constant of integration to be determined à posteriori.

Wall layer

To describe the flow behavior near the wall we introduce the inner variable

$$\hat{y} = y\hat{\varepsilon} = y\varepsilon R_w\,,$$

where $R_w = U_\infty L/\nu_w$, ν_w = local kinematic viscosity at the wall. The appropriate asymptotic expansions for this region are

$$\hat{u} = \varepsilon \hat{u}_1(x, \hat{y}) + \varepsilon^2 \hat{u}_2(x, \hat{y}) + \dots\,, \tag{23a}$$

$$\hat{v} = \varepsilon\hat{\varepsilon}[\hat{v}_1(x, \hat{y}) + \varepsilon \hat{v}_2(x, \hat{y}) + \dots]\,, \tag{23b}$$

$$\hat{p} = \hat{p}_1(x, \hat{y}) + \varepsilon \hat{p}_2(x, \hat{y}) + \dots\,, \tag{23c}$$

$$\hat{\varrho} = \hat{\varrho}_1(x, \hat{y}) + \varepsilon \hat{\varrho}_2(x, \hat{y}) + \dots\,, \tag{23d}$$

$$\hat{t} = \hat{t}_1(x, \hat{y}) + \varepsilon \hat{t}_2(x, \hat{y}) + \dots\,, \tag{23e}$$

$$\hat{\tau}_{ab} = \varepsilon^2 \hat{\tau}_{ab_1}(x, \hat{y}) + \varepsilon^3 \hat{\tau}_{ab_2}(x, \hat{y}) + \dots\,, \tag{23f}$$

$$\hat{\tau}_{abc} = \varepsilon^3 \hat{\tau}_{abc_1}(x, \hat{y}) + \varepsilon^4 \hat{\tau}_{abc_2}(x, \hat{y}) + \dots\,. \tag{23g}$$

An asymptotic expansion for the viscosity is derived expanding μ in a Taylor series around t_1, that is,

$$\mu = \mu(\hat{t}_1) + \varepsilon \mu_t(\hat{t}_1)\, \hat{t}_2 + O(\varepsilon^2)\,, \tag{24}$$

where, $\mu_t = \left.\dfrac{\partial \mu}{\partial t}\right|_{\hat{t}=\hat{t}_1}$.

The result of inserting (23) and (24) into (1—6) and collecting terms of same order is:

to lowest order terms:

$$(\hat{\varrho}_1 \hat{u}_{j_1})_{x_j} = 0\,, \tag{25a}$$

$$(\hat{\varrho}_1 \hat{\tau}_{uv_1})_{\hat{y}} + \frac{\mu(\hat{t}_1)}{\nu_w}\, \hat{u}_{1_{\hat{y}\hat{y}}} = 0\,, \tag{25b}$$

$$\hat{p}_{1_{\hat{y}}} = 0\,, \qquad \hat{t}_{1_{\hat{y}\hat{y}}} = 0\,, \tag{25 c, 25 d}$$

$$\hat{p}_1 = \frac{\gamma - 1}{\gamma}\, \frac{1}{D}\, \hat{\varrho}_1 \hat{t}_1; \tag{25e}$$

to first order terms:

$$(\hat{\varrho}_1 \hat{u}_{j_2})_{x_j} + (\hat{\varrho}_2 \hat{u}_{j_1})_{\hat{x}_j} = 0\,, \tag{26a}$$

$$(\hat{\varrho}_1 \hat{\tau}_{uv_2})_{\hat{y}} + (\hat{\varrho}_2 \hat{\tau}_{uv_1})_{\hat{y}} + \hat{\tau}_{\varrho uv_{1_{\hat{y}}}} + \frac{1}{\nu_w} [\mu(\hat{t}_1)\, \hat{u}_{2_{\hat{y}\hat{y}}} + \mu_t(\hat{t}_1)\, \hat{t}_2 \hat{u}_{1_{\hat{y}\hat{y}}}] = 0\,, \tag{26b}$$

$$p_{2_{\hat{y}}} = 0\,, \tag{26c}$$

$$\frac{1}{\sigma \nu_w} [\mu(\hat{t}_1)\, \hat{t}_{2_{\hat{y}\hat{y}}}] + (\hat{\varrho}_1 \hat{\tau}_{tv_1})_{\hat{y}} = 0\,, \tag{26d}$$

$$\hat{p}_2 = \frac{\gamma - 1}{\gamma}\, \frac{1}{D}\, (\hat{\varrho}_1 \hat{t}_2 + \hat{\varrho}_2 \hat{t}_1)\,. \tag{26e}$$

Equations 25 (c—e) and the matching conditions show that pressure, temperature and density are constant across the wall layer up to the lowest order terms. Hence, we can write

$$\left.\begin{aligned} p_1 &= p_1(x, y) = P_{10}\,, \\ \hat{t}_1 &= t_W\,, \quad \hat{\varrho}_1 = \varrho_W\,. \end{aligned}\right\} \tag{27}$$

Furthermore, for an impermeable wall, the matching of the pressure in the wall, outer and inviscid regions gives

$$\hat{p}_2(x, \hat{y}) = p_2(x, y) = P_2(x, Y) = 0\,, \tag{28}$$

The velocity profiles for the defect and wall layers are given by

$$u = U_{10} + \varepsilon(a_{11} \ln y + a_{12}) + \varepsilon^2(a_{21} \ln^2 y + a_{22} \ln y + a_{23}) + O(\varepsilon^3) \tag{29a}$$

and

$$\hat{u} = \varepsilon(\hat{a}_{11} \ln \hat{y} + \hat{a}_{12}) + \varepsilon^2(\hat{a}_{21} \ln^2 \hat{y} + \hat{a}_{22} \ln \hat{y} + \hat{a}_{23}) + O(\varepsilon^3) \tag{29b}$$

where the a's represent arbitrary functions of y.

The presence of first order logarithmic terms in the above expressions is justified by equation (25b) and the matching conditions for the defect and wall layers [AFZAL, 2]. The second-order bilogarithmic terms occur due to expression (26b). Indeed, equations (26d) and (26e) yield that the first-order density term is logarithmic; substitution of this term and the solution of (25b) into (26b) results in a solution which contains bilogarithmic terms. Thus bilogarithmic terms occur only in compressible flow.

An analysis similar to that for the velocity profiles carried out for the temperature profiles, suggests that

$$t = t_{10} + \varepsilon(b_{11} \ln y + b_{12}) + \varepsilon^2(b_{21} \ln^2 y + b_{22} \ln y + b_{23}) + O(\varepsilon^3)\,, \tag{30a}$$

and

$$\hat{t} = t_w + \varepsilon(\hat{b}_{11} \ln \hat{y} + \hat{b}_{12}) + \varepsilon^2(\hat{b}_{21} \ln^2 \hat{y} + \hat{b}_{22} \ln \hat{y} + \hat{b}_{23}) + O(\varepsilon^3)\,, \tag{30b}$$

where the b's are arbitrary functions of y.

On substituting the above profiles into the equation of state and taking into account the conditions imposed by equations (27) and (28), it follows that the density profiles are given by

$$\varrho = P_{10} - \varepsilon \frac{P_{10}}{T_{10}}(b_{11} \ln y + b_{12}) + \varepsilon^2(c_{21} \ln^2 y + c_{22} \ln y + c_{23}) + O(\varepsilon^3)\,, \tag{31a}$$

and

$$\hat{\varrho} = \varrho_W - \varepsilon \frac{\varrho_W}{t_W}(\hat{b}_{11} \ln \hat{y} + \hat{b}_{12}) + \varepsilon^2(\hat{c}_{21} \ln^2 \hat{y} + \hat{c}_{22} \ln \hat{y} + \hat{c}_{23}) + O(\varepsilon^3)\,. \tag{31b}$$

where the c's are arbitrary functions of y.

Now, whatever techniques are used, matching is always carried out by comparing the outer expansion for small y with the inner expansion for large $\hat{y}$; the limit of the outer expansions as $y \to 0$, and the limit of the inner expansion as $\hat{y} \to \infty$, are shown in Table 1.

In this Table, each row defines a set of three equations to be satisfied, the tangential velocity, the temperature and the density matching condition equations. It is interesting to note that as $\hat{y} \to \infty$, the logarithmic terms give contributions of lower order of magnitude than the order formally indicated; this fact is fundamental for a successful match. Indeed, if the bilogarithmic terms are dropped, the matching condition between the velocity profiles yields

$$U_{10} = \varepsilon \ln \hat{\varepsilon}\, \hat{a}_{11} + \varepsilon \hat{a}_{12} - \varepsilon a_{12} \tag{32a}$$

so that $\varepsilon \ln \hat{\varepsilon} = O(1)$ as $\varepsilon \to 0$.

Equation (32) is the classical skin-friction equation obtained by MELLOR [12]. Thus the temperature and density matching conditions become

$$t_W + \varepsilon \ln \hat{\varepsilon} \hat{b}_{11} + \varepsilon \hat{b}_{12} = T_{10} + \varepsilon b_{12} \tag{33a}$$

and

$$\varrho_W + \varepsilon \ln \hat{\varepsilon} \frac{\varrho_W}{t_W} \hat{b}_{11} + \varepsilon \frac{\varrho_W}{t_W} \hat{b}_{12} = P_{10} + \varepsilon \frac{P_{10}}{T_{10}} b_{12}\,. \tag{34a}$$

The matching difficulties can now be clearly seen from equations (33a) and (34a), that is, as $\varepsilon \to 0$ parameter $\hat{b}_{11}$ alone is not capable of satisfying both equations. Hence, in order to make the matching feasible, one must consider the second-order bilogarithmic terms. Then the matching conditions given by equations (32a—34a) become

$$U_{10} = \varepsilon \ln \hat{\varepsilon} \hat{a}_{11} + \varepsilon \hat{a}_{12} - \varepsilon a_{12} + \varepsilon^2 \ln^2 \hat{\varepsilon} \hat{a}_{21} + \varepsilon^2 \ln \hat{\varepsilon} \hat{a}_{21} + \varepsilon^2 a_{23} - \varepsilon^2 \hat{a}_{23}\,, \tag{32b}$$

$$t_W - T_{10} = -\varepsilon \ln \hat{\varepsilon} \hat{b}_{11} - \varepsilon \hat{b}_{12} + \varepsilon b_{12} - \varepsilon^2 \ln^2 \hat{\varepsilon} \hat{b}_{21} - \varepsilon^2 \ln \hat{\varepsilon} \hat{b}_{22} - \varepsilon^2 \hat{b}_{23} + \varepsilon^2 b_{23}\,, \tag{33b}$$

$$P_{10} - \varrho_W = \varepsilon \ln \hat{\varepsilon} \frac{\varrho_W}{t_W} \hat{b}_{11} + \varepsilon \frac{\varrho_W}{t_W} \hat{b}_{12} - \varepsilon \frac{P_{10}}{T_{10}} b_{12} + \varepsilon^2 \ln^2 \hat{\varepsilon} c_{21} + \varepsilon \ln \hat{\varepsilon} \hat{c}_{22} - \varepsilon^2 c_{23} + \varepsilon^2 \hat{c}_{23}\,. \tag{34b}$$

The system of equations defined above has now a solution since its number of parameters is greater than the number of equations. The trick is obviously done by the second order bilogarithmic terms which become order unity in the overlap region.

In AFZAL's analysis, it was not clear that solutions of the form (29a) to (31b) could satisfy the matching conditions since the density profiles were omitted. Furthermore, in that analysis the bilogarithmic terms were obtained by assuming that the Reynolds stress terms could be expanded in a particular form, whereas here these bilogarithmic terms arise from the solution of the equations of motion.

As shown by the equations of motion of second order, the assumption that second-order bilogarithmic terms occur for compressible flow is reasonable and this is a direct consequence of the fact that the first order approximation solution contains

Table 1. Matching conditions

			zeroth order	first order		second order		
tangential velocity	outer exp ($y \to 0$)	$u =$	U_{10}	$+\varepsilon a_{11} \ln y$	$+\varepsilon a_{12}$	$+\varepsilon^2 a_{21} \ln^2 y$	$+\varepsilon^2 a_{22} \ln y$	$+\varepsilon^2 a_{23}$
	inner exp ($\hat{y} \to \infty$)	$\hat{u} =$	$+\varepsilon \ln \hat{\varepsilon} \hat{a}_{11}$ $+\varepsilon^2 \ln^2 \hat{\varepsilon} \hat{a}_{21}$	$+\varepsilon \hat{a}_{11} \ln y$ $+2\varepsilon^2 \ln \hat{\varepsilon} \hat{a}_{21} \ln y$	$+\varepsilon \hat{a}_{12}$ $+\varepsilon^2 \ln \hat{\varepsilon} \hat{a}_{22}$	$+\varepsilon^2 \hat{a}_{21} \ln^2 y$	$+\varepsilon^2 \hat{a}_{22} \ln y$	$+\varepsilon^2 \hat{a}_{23}$
temperature	outer exp ($y \to 0$)	$t =$	T_{10}	$+\varepsilon b_{11} \ln y$	$+\varepsilon b_{12}$	$+\varepsilon^2 b_{21} \ln^2 y$	$+\varepsilon^2 b_{22} \ln y$	$+\varepsilon^2 b_{23}$
	inner exp ($\hat{y} \to \infty$)	$\hat{t} =$	t_w $+ \varepsilon \ln \hat{\varepsilon} \hat{b}_{11}$ $+ \varepsilon^2 \ln^2 \hat{\varepsilon} \hat{b}_{21}$	$+\varepsilon \hat{b}_{11} \ln y$ $+2\varepsilon^2 \ln \hat{\varepsilon} \hat{b}_{20} \ln y$	$+\varepsilon \hat{b}_{12}$ $+\varepsilon^2 \ln \hat{\varepsilon} \hat{b}_{22}$	$+\varepsilon^2 \hat{b}_{21} \ln^2 y$	$+\varepsilon^2 b_{22} \ln y$	$+\varepsilon^2 \hat{b}_{23}$
density	outer exp ($y \to 0$)	$\varrho =$	P_{10}	$-\varepsilon \frac{P_{10}}{T_{10}} \hat{b}_{11} \ln y$	$-\varepsilon \frac{P_{10}}{T_{10}} \hat{b}_{12}$	$+\varepsilon^2 \hat{c}_{21} \ln^2 y$	$+\varepsilon^2 \hat{c}_{22} \ln y$	$+\varepsilon^2 \hat{c}_{23}$
	inner exp ($\hat{y} \to \infty$)	$\hat{\varrho} =$	ϱ_w $+\varepsilon \ln \hat{\varepsilon} \frac{\varrho_w}{t_w} \hat{b}_{11}$ $+\varepsilon^2 \ln^2 \hat{\varepsilon} \hat{c}_{21}$	$-\varepsilon \frac{\varrho_w}{t_w} \hat{b}_{11} \ln y$ $+ 2\varepsilon^2 \ln \hat{\varepsilon} \hat{c}_{21} \ln y$	$-\varepsilon \frac{\varrho_w}{t_w} \hat{b}_{12}$ $+\varepsilon^2 \ln \hat{\varepsilon} \hat{c}_{22}$	$+\varepsilon^2 \hat{c}_{21} \ln^2 y$	$+\varepsilon^2 \hat{c}_{22} \ln y$	$+\varepsilon^2 \tilde{c}_{23}$

logarithmic terms. Higher order asymptotic theories should then include such terms.

V. Conclusion

In this work we have shown that application to compressible turbulent boundary layers of the intermediate variable technique and of the matched asymptotic expansion method indicates that the defect and the wall layer solutions have an overlap domain.

The arguments of other authors based purely on considerations of the velocity profiles are not enough to settle the matching question. Indeed, if no further comments about the nature of the density expressions were necessary the following simple analysis would apply. In essence, what one has to show is that the difference between the outer and the inner approximations converges uniformly to zero in an overlap domain. In other words, one has to show that for each $\delta > 0$ there exists an $\varepsilon_\delta > 0$, such that if $\varepsilon < \varepsilon_\delta$ and $\mu(\varepsilon) \leqq x \leqq \nu(\varepsilon)$, $\mu(\varepsilon)$ and $\nu(\varepsilon)$ two class function defined in D_{overlap}, then $|w(x, \varepsilon)| = |u_{\text{inner}} - u_{\text{outer}}| < \delta$.

Now assuming that for the incompressible case there exists an overlap domain, it follows that

$$\forall \delta' > 0 , \quad \exists \varepsilon'_\delta > 0 , \varepsilon < \varepsilon'_\delta \wedge \mu(\varepsilon) \leqq x \leqq \nu(\varepsilon) \to \to |u_{i_{\text{inner}}} - u_{i_{\text{outer}}}| < \delta' .$$

This result and the fact that sine is a Lipschitz function, that is there exists a $K > 0$ (the constant of Lipschitz) such that, for each $x, y \in R$, we have $|\sin x - \sin y| < K|x - y|$ (for example, take $K = 1$), imply that if for the compressible case we take $\varepsilon_\delta = \varepsilon'_\delta$ and $\delta = \delta' K$, then for $\varepsilon < \varepsilon_\delta$ and $\mu(\varepsilon) \leqq x \leqq \nu(\varepsilon)$, μ and ν the same class of function above, we have

$$|u_{\text{inner}} - u_{\text{outer}}| = |\sin u_{i_{\text{inner}}} - \sin u_{i_{\text{outer}}}| < < K|u_{i_{\text{inner}}} - u_{i_{\text{outer}}}| < K\delta' < \varepsilon .$$

This would complete the proof that for the compressible case the inner and the outer expansion have an overlap domain which contains the overlap domain of the incompressible case. Indeed, this is also indicated by the analysis using the intermediate variable technique.

Acknowledgement

In the course of this research the author benefited from many useful discussions with Dr. L. C. Squire. Valuable comments were also received from Mr. N. Chokani. The work was financially supported partly by the Conselho Nacional de Desenvolvimento Cientifico e Technologico — CNPq (Brazilian National Research Council), and partly by an ORS award.

References

1 Adamson, T. C. Jr.; Feo, A.: Interaction between a shock wave and a turbulent boundary layer at transonic speeds. SIAM J. Appl. Math. 29 (1975), 121–145.
2 Afzal, N.: A higher order theory for compressible turbulent boundary layers at moderately large Reynolds number. J. Fluid Mech. 57 (1973), 1–27.
3 Bush, W. B.; Fendell, F.: Asymptotic analysis of turbulent channel and boundary layer flow. J. Fluid Mech. 56 (1972), 657–681.
4 Cebeci, T.; Smith, A. O. M.: Analysis of turbulent boundary layers. Academic Press, 1974.
5 Kaplun, S.: Fluid mechanics and singular perturbations. Academic Press, 1967.
6 Kistler, A. L.: Fluctuation measurements in a supersonic turbulent boundary layer. Phys. Fluids 2 (1959), 290–296.
7 Kistler, A. L.; Chen, W. S.: A fluctuating pressure field in a supersonic turbulent boundary layer. J. Fluid Mech. 16 (1963), 41–64.
8 Lagerstrom, P. A.; Casten, R. G.: Basic concepts underlying singular perturbation techniques. SIAM Review 14 (1972), 63–120.
9 Liepmann, H. W.; Roscko, A.; Elements of gasdynamics. John Wiley & Sons, 1957.
10 Liou, M. S.; Adamsom, T. C. Jr.: Interaction between a normal shock wave and a turbulent boundary layer at high transonic speeds. Part II: Wall shear stress. ZAMP 31 (1980), 227–246.
11 Maise, G.; McDonald, H.: Mixing length and kinematic eddy viscosity in compressible boundary layers. AIAA J. 6 (1968), 73–80.
12 Mellor, G. L.: The large Reynolds number asymptotic theory of turbulent boundary layers. Int. J. Engng. Sci. 10 (1972), 851–873.
13 Melnik, R. E.; Grossman, B.: Analysis of the interaction of a weak normal shock wave with a turbulent boundary layer. AIAA Paper No 74–598 (1974).
14 Morkovin, M. V.: Effects of compressibility on turbulent flows. In: Mécanique de la turbulence. 1962, pp. 367–380.
15 Yajnik, K.: Asymptotic theory of turbulent shear flow. J. Fluid Mech. 42 (1970), 411–427.

Received October 19, 1987, revised version February 5, 1988

Address: Dr. A. P. Silva Freire, Engineering Department, Cambridge University, Cambridge CB2 1PZ, England, presently at Programa de Engenharia Mecanica (COPPE/UFRJ), Universidade Federal do Rio de Janeiro, Brazil

ZAMM · Z. angew. Math. Mech. 69 (1989) 2, 104–106

Atanacković, T. M.; Djukić, Dj. S.; Strauss, A. M.

A Variational Principle for an Isothermal Chemical Reaction

1. Introduction

Let A be a substance contained in a gas that flows through a cylindrical pore of a catalyst. On the walls of the pore the substance A transforms into another substance, say B. We assume that the transformation of A into B is an isothermal chemical reaction of the first order [1]. Then, after several changes of variables, the equation describing the dimensionless concentration of the substance A along the (dimensionless) length t of the catalyst pore can be written as [2]

$$\frac{\mathrm{d}}{\mathrm{d}t}(a\dot{q}) + \lambda \frac{1 - q}{1 + \nu(1 - q)} = 0 , \qquad t \in (0, 1) , \tag{1}$$

where $(.)^{\cdot} = \mathrm{d}/\mathrm{d}t\,(.)$, q is the dimensionless concentration, $a(t) > 0$ is the dimensionless diffusion coefficient and $\lambda \geqq 0$ and $\nu > 0$ are coefficients depending on temperature and initial concentration, respectively.

The boundary conditions corresponding to (1) are

$$q(0) = q(1) = 0 . \tag{2}$$

The boundary value problem (1), (2) has been treated numerically in [2]. Also in [2] it is shown that (1), (2) has for any $\lambda \geqq 0$ a unique $C^2[0, 1]$ solution and that the following bounds hold [2 p. 53]:

$$\begin{aligned} &0 < q(t) < 1 \quad \text{for} \quad t \in (0, 1) , \qquad \lambda > 0 , \\ &q(t) \equiv 0 \quad \text{for} \quad \lambda = 0 . \end{aligned} \tag{3}$$

Our intention in this note is to treat the boundary value problem (1), (2) by a variational method developed in [3]. First, we shall show that the functional constructed by the procedure described in [3] attains a global minimum on the solution of (1), (2). Then, we shall use the Ritz method to determine approximate solutions to (1), (2) for several values of the parameters λ and ν. Finally, we shall present error estimates for the approximate solutions. The error estimates will be based on the values of the functional calculated on approximate solutions.

2. Variational principle and error estimate

To escort the procedure from [3] we first note that (1), (2) is equivalent to the stationarity condition ($\delta I_1 = 0$) for the functional I_1 given by

$$I_1 = \int_0^1 L \,\mathrm{d}t , \qquad L = \frac{a}{2}\dot{q}^2 - \frac{\lambda}{\nu} q - \frac{\lambda}{\nu^2} \ln [1 + \nu(1 - q)] . \tag{4}$$

By using L, given by $(4)_2$, we can follow the procedure of [3] to obtain the following functional

$$I = \int_0^1 \left\{ a\dot{Q}^2 - \frac{\lambda}{\nu} Q - \frac{\lambda}{\nu^2} \ln [1 + \nu(1 - Q)] + \frac{\lambda}{\nu^2} \ln \frac{\lambda}{\lambda + \nu(\dot{a}\dot{Q} + a\ddot{Q})} + \frac{\lambda + (1 + \nu)(\dot{a}\dot{Q} + a\ddot{Q})}{\nu} \right\} \mathrm{d}t . \tag{5}$$

In (5) $Q(t)$ is an admissible trial function, that is, $Q \in W$, where

$$W = \left\{Q: Q \in C^2[0, 1]; Q(0) = Q(1) = 0; 0 < Q(t) < 1, \right.$$
$$\left. \frac{\mathrm{d}}{\mathrm{d}t}(a\dot{Q}) \leqq 0\right\}. \tag{6}$$

In [3] we showed that (5) is stationary and satisfies a necessary extremality condition (second variation positive definite) on the exact solution q of (1), (2). Also the value of I on q is equal to zero so that

$$I(q) = 0; \quad \delta I(q,f) = 0; \quad \delta^2 I(q,f) > 0\,, \tag{7}$$

where $f = Q - q$ and $Q \in W$. Relations (7) could be, of course, checked directly from (5) by calculating the first and second variations and using (1) and (2) together with partial integration.

We now show that q is the global minimizer of I on W. Let $Q \in W$, then, expanding (5), we get

$$I(Q) = I(q) + \delta I(q,f) + \tfrac{1}{2}\,\delta^2 I(\Psi,f)\,, \tag{8}$$

where

$$\Psi = q + \varepsilon f = q + \varepsilon(Q - q)\,, \qquad 0 < \varepsilon < 1\,. \tag{9}$$

Using $(7)_{1,2}$ in (8) and calculating the second variation of (5) we get

$$I(Q) = \frac{1}{2}\int_0^1 \left\{\frac{\lambda}{[1 + \nu(1 - \Psi)]^2}\, f^2 + 2a\dot{f}^2 + \right.$$
$$\left. + \frac{\lambda}{[\lambda + \nu(\dot{a}\dot{\Psi} + a\ddot{\Psi})]^2}\,(\dot{a}\dot{f} + a\ddot{f})^2\right\}\mathrm{d}t\,. \tag{10}$$

From (9) and the propeties of Q and q we conclude that

$$0 < \Psi < 1; \quad \frac{\mathrm{d}}{\mathrm{d}t}(a\dot{\psi}) \leqq 0\,. \tag{11}$$

Using (11) we get

$$\frac{\lambda}{[1 + \nu(1 - \Psi)]^2} \geqq \frac{\lambda}{(1 + \nu)^2}\,; \qquad \frac{\lambda}{[\lambda + \nu(\dot{a}\dot{\Psi} + a\ddot{\Psi})]^2} \geqq \frac{1}{\lambda}\,, \tag{12}$$

so that (10) becomes

$$I(Q) \geqq \frac{1}{2}\int_0^1 \left\{\frac{\lambda}{(1 + \nu)^2}\, f^2 + 2a\dot{f}^2 + \frac{1}{\lambda}\,(\dot{a}\dot{f} + a\ddot{f})^2\right\}\mathrm{d}t\,. \tag{13}$$

Let Δ be a constant such that

$$\int_0^1 \left\{\frac{\lambda}{(1 + \nu)^2}\, h^2 + \frac{1}{\lambda}\,(\dot{a}\dot{h} + a\ddot{h})^2\right\}\mathrm{d}t \geqq \Delta \int_0^1 \dot{h}^2\,\mathrm{d}t \tag{14}$$

for all $h \in C^2[0, 1]$ with $h(0) = h(1) = 0$. The values of Δ, for specific $a(t)$ (for example $a(t) = \text{const.} > 0$) and different boundary conditions on h, could be found in [4]. In the next section we shall determine the best constant in (14) using the methods of optimal control theory [5], [6]. With known Δ and setting $h = f = Q - q$ the inequality (13) becomes

$$I(Q) \geqq \tfrac{1}{2}\,(\Delta + 2a_{\min})\,\|\dot{f}\|^2_{L_2}\,, \tag{15}$$

where $a_{\min} = \inf_{t \in [0,1]} a(t)$ and $\|\dot{f}\|_{L_2} = (\int_0^1 \dot{f}^2\,\mathrm{d}t)^{1/2}$ is the L_2-norm of $\dot{f}$. The boundary conditions and the Cauchy-Schwartz inequality lead to

$$\|f\|_{L_\infty} = \sup_{t \in [0,1]} |f(t)| \leqq \tfrac{1}{2}\,\|\dot{f}\|_{L_2}\,, \tag{16}$$

so that (15) transforms to

$$I(Q) \geqq 2(\Delta + a_{\min})\,\|f\|^2_{L_\infty} = 2(\Delta + a_{\min})\,\|Q - q\|^2_{L_\infty}\,. \tag{17}$$

Inequality (17) represents the sufficient condition for the existence of a minimum of I on the set W [7]. Since (17) is valid for any $Q \in W$, we can state the following

Theorem: *The exact solution q to (1), (2) is the global minimizer of I given by (5) on the set W defined by (6) with $I(q) = 0$.*

In applications $Q(t)$ will be an approximate solution to (1), (2). Then, from (17) we have an error estimate of such an approximate solution as

$$\|Q - q\|_{L_\infty} \leqq \left\{\frac{I(Q)}{2(\Delta + 2a_{\min})}\right\}^{1/2}\,. \tag{18}$$

3. Application

We now apply the results obtained in the previous section to a specific example. Following [2] we choose the following values for p and ν:

$$p(t) = 1\,, \qquad \nu = 1\,. \tag{19}$$

Equation (1) then becomes

$$\ddot{q} + \lambda\frac{1 - q}{2 - q} = 0\,. \tag{20}$$

We shall first determine the largest value for Δ in the inequality (14) which now takes the form

$$\int_0^1 \left(\frac{\lambda}{4}\, h^2 + \frac{1}{\lambda}\,\ddot{h}^2\right)\mathrm{d}t \geqq \Delta \int_0^1 \dot{h}^2\,\mathrm{d}t\,. \tag{21}$$

Consider the problem of determining the minimum of the functional K defined as

$$K = \left(\int_0^1 \left(\frac{\lambda}{4}\, h^2 + \frac{1}{\lambda}\,\ddot{h}^2\right)\mathrm{d}t\right)\Big/\left(\int_0^1 \dot{h}^2\,\mathrm{d}t\right) \tag{22}$$

among all $h \in C^2[0, 1]$ with $h(0) = h(1) = 0$.

Following [5] and [6] we transform (22) to the problem: Minimize

$$K = \left(\int_0^1 \left(\frac{\lambda}{4}\, x_1^2 + \frac{1}{\lambda}\, u^2\right)\mathrm{d}t\right)\Big/\left(\int_0^1 x_2^2\,\mathrm{d}t\right) \tag{23}$$

subject to the differential constraints

$$\dot{x}_1 = x_2\,, \qquad \dot{x}_2 = u\,, \tag{24}$$

with $x_1(0) = x_1(1) = 0$. In (24) u is the "control". The optimal control problem (23), (24) is, due to the form of (23), a nonstandard control problem. It was shown, however, in [4] that the minimum of K could be determined using (a version of) Pontryagin's maximum principle. Applying the procedure given in [6], $\min K = \Delta$ is determined as

$$\Delta = \min_n \left[\frac{\lambda}{4(n\pi)^2} + \frac{(n\pi)^2}{\lambda}\right], \qquad n = 1, 2, 3, \ldots. \tag{25}$$

For different λ the minimum on the right hand side of (25) is attained at different n. In Table 1 we listed Δ for few values of λ.

Table 1

λ	10	20	30	100	300
Δ	1.24	1.00	1.08	1.028	1.001

To determine the approximate solution to (20), (2), we take the trial function in the form

$$Q = \sum_{i=1}^{3} c_i t^i (1 - t)^i\,, \tag{26}$$

where c_i, $i = 1, 2, 3$, are constants. The values of c_i are determined by substituting (26) into (see (5))

$$I = \int_0^1 \left[\dot{Q}^2 - \lambda Q - \lambda \ln(2 - Q) + \lambda \ln \frac{\lambda}{\lambda + \ddot{Q}} + \lambda + 2\ddot{Q}\right]\mathrm{d}t \tag{27}$$

and minimizing the resulting expression. The results, together with the error estimates obtained from (18), are presented in Table 2.

Table 2

λ	c_1	c_2	c_3	I	$\leqq \|\|f\|\|_{L_\infty}$
10	32.16	−0.51	−0.05	$4.20 \cdot 10^{-5}$	$2.54 \cdot 10^{-3}$
20	52.52	−1.73	−0.61	$3.26 \cdot 10^{-5}$	$2.33 \cdot 10^{-3}$
30	67.00	−3.34	−1.25	$1.38 \cdot 10^{-6}$	$1.49 \cdot 10^{-3}$
100	118.72	−14.2	−0.25	$9.30 \cdot 10^{-2}$	$1.23 \cdot 10^{-1}$
300	206.08	84.17	51.52	$11.2 \cdot 10^{-2}$	$1.36 \cdot 10^{-1}$

From Table 2 we conclude that, although simple, the approximate solution of the form (26) has remarkable accuracy for moderate values of λ.

Acknowledgement: This research was supported in part by U.S. NSF JFP 524 and SIZ NR Vojvodine.

References

1 Bird, R. B.; Stewart, N. B.; Lighfoot, E. N.: Transport phenomena. John Wiley, New York 1960.
2 Bohl, E.: Finite Modelle gewöhnlicher Randwertaufgaben. Teubner, Stuttgart 1981.
3 Djukić, Dj. S.; Atanacković, T. M.; J. Math. Anal. Appl. 75 (1980), 203 to 218.
4 Mitrinović, D. S., Analytic Inequalities. Springer, Berlin 1970.
5 Subrahmanyan, M. B.: J. Math. Anal. Appl. 60 (1977), 601–616.
6 Subrahmanyan, M. B.: J. Math. Anal. Appl. 77 (1980), 47–59.
7 Segan, H.: Introduction to the calculus of variations, McGraw-Hill 1969.

Received July 1, 1987

Addresses: Dr. T. M. Atanacković, Dr. Djordje Djukić, Faculty of Technical Sciences, University of Novi Sad, YU-21000 Novi Sad, Yugoslavia; Dr. A. M. Strauss, Vanderbilt University, Nashville, Tennessee 37235, U.S.A.

ZAMM · Z. angew. Math. Mech. 69 (1989) 2, 106–109

Schubert, M.

The Evaluation of Integrals and Derivatives by Means of Physically Implied Interpolation Functions with a Finite Limiting Fourier Frequency

Frequently objective reasons suggest the use of interpolation functions with a finite limiting Fourier frequency. The results relating to this are compared with those of standard interpolation methods concerning the strength of interpolation vibrations and of the mean summation error. The physically implied function reveals superiority compared with the Lagrange and spline procedures.

1. Introduction

The evaluation of interpolation values, derivatives, integrals from a set of measured or given ordinates $y(x_n)$, which are assumed to characterize a bounded real function $y(x)$ at the discrete values $x_0, \dots, x_n, \dots, x_m$ of the variable x, plays an important role in applied mathematics and physical measuring technique. The result of this evaluation depends qualitatively and quantitatively on the general properties of the function used for interpolation. As a rule, a transformation of the variable x can be applied in such a way that the differences $(x_{n+1} - x_n)$ can be assumed to be constant.

Important well-known types of interpolation functions are the *Lagrange polynomial* $L(x)$ and the *spline function* $C(x)$ (with cubic polynomials in the subintervals). In the present paper in Section 2 will be shown that in physics and technics frequently objective reasons suggest the use of interpolation functions with a finite limiting Fourier frequency $\hat{f}$. This property leads to a specific type $S(x)$ of a function, which we will call *physically implied* (PI) *interpolation function*. Independently of the physical and technical arguments on the specific form of $S(x)$ we shall later on compare the mathematical properties of $S(x)$ with those of the functions $L(x)$ and $C(x)$. In doing this we shall consider the derivatives and the strength of interpolation vibrations in Section 3 and the integration and summation errors in Section 4, where we in particular shall focus our attention to the generality of assumptions and a wide range of validity as well as to the applicability of the remainder terms. Altogether there results that the PI-function $S(x)$ has more advantageous properties than the Lagrange and spline functions have [1, 2].

Finally we will compile the basic relations assigned to the Lagrange and spline interpolation functions. This compilation will enable us to make reference to $L(x)$ and $C(x)$ without interrupting the comparison between the various types of interpolation functions.

The *Lagrange function* can be written as

$$L(x) = \sum_{n=0}^{m} y(n)\, l_{n,m}(x) \quad \text{with} \quad l_{n,m}(n') = \begin{cases} 1 & \text{for} \quad n' = n \\ 0 & \text{for} \quad \text{other } n' \end{cases}. \tag{1.1}$$

The "expansion functions" $l_{n,m}(x)$ are polynomials of m-th degree.

The *spline function* [3] is given by

$$C(x) = c_{n,m}(x) \quad \text{for} \quad x \in [n, n+1] \quad \text{with} \quad n = 0, \dots, m-1\,. \tag{1.2}$$

$$c_{n,m}(x) = y(n) + [y(n+1) - y(n) - \tfrac{1}{3}\, C''(n+1) - \tfrac{1}{6}\, C''(n)] \times \\ \times (x - n) + \tfrac{1}{2}\, C''(n)\,(x-n)^2 + \\ + [\tfrac{1}{6}\, C''(n+1) - \tfrac{1}{6}\, C''(n)](x-n)^3\,.$$

The second derivatives $C''(n)$ are determined in such a way that $C'(x)$ and $C''(x)$ exist uniquely in $[0, m]$. These interpolation conditions yield for $(m-1)$ quantities $C''(n)$, say for $C''(1), \dots, C''(m-1)$, the $(m-1)$ equations

$$C''(n-1) + 4C''(n) + C''(n+1) = 6y(n-1) - 12y(n) + \\ + 6y(n+1) \tag{1.3}$$

with $n = 1, \dots, m-1$. Let us choose $C''(0) = 0$, $C''(m) = 0$ for the two quantities undetermined so far; this is a choice in favour of the spline function concerning the comparisons to be performed in Section 3. The function $C(x)$ is an element of the normalized linear space $C^2(0, m)$ and it can be shown that the total curvature attains a minimum value:

$$\int_0^m \mathrm{d}x\, |C''(x)|^2 = \text{Min}\,. \tag{1.4}$$

2. Basic properties of the physically implied interpolation function $S(x)$

The occurrence of a finite limiting Fourier frequency in physical and technical problems follows, as a rule, the same pattern as in the representative case of optical imaging. There the relation

$$y_{\text{out}}(x) = \int_{-\infty}^{+\infty} \mathrm{d}x'\, y_{\text{in}}(x')\, F(x - x') \tag{2.1}$$

expresses the connection between the input signal y_{in} and the output (image) signal y_{out} as a function of the position coordinate. F is the instrumental (filtering) function. For an ideal device $F(x - x')$ has the character of a delta function (with a constant Fourier spectrum extending to infinity). Due to the, in reality, always existing finite optical resolving power the instrumental function $F(x - x')$ is nonzero within a certain non-vanishing interval and tends to zero if $|x - x'|$ tends to infinity; its Fourier transform $\boldsymbol{F}(f)$ has an effective finite limiting Fourier frequency $\hat{f}$. Equation (2.1) leads to the relation

$$\boldsymbol{y}_{\text{out}}(f) = \boldsymbol{y}_{\text{in}}(f) \cdot \boldsymbol{F}(f) \tag{2.2}$$

for the Fourier transforms. This reveals that the properties of the instrumental function imply, under realistic conditions for $\boldsymbol{y}_{\text{in}}(f)$, a finite limiting frequency for $\boldsymbol{y}_{\text{out}}(f)$ as well. The same pattern holds for the output-input-relation of a linear electric four-terminal network.

These examples exhibit that important functions $S(x)$ in physics and technics can be characterized by the following properties (that we shall call PI-properties):

$S(x)$ is bounded and Fourier-transformable . (2.3)

The (improper) *integrals* $\int_{-\infty}^{+\infty} dx\, S(x)^2 = \int_{-\infty}^{+\infty} df |\mathbf{S}(f)|^2$ *exist .*

The PI-properties render it possible to assign to $\mathbf{S}(f)$ an effective limiting frequency $\hat{f}$, so that $|\mathbf{S}(f)| = 0$ for $|f| > \hat{f}$. The term "effective" means that the improper integral over f can be replaced in good approximation by an integral over a finite f-interval $[-\hat{f}, \hat{f}]$, where $\hat{f}$ must be chosen according to the measuring situation and/or the required accuracy. Correspondingly, an effective finite sampling point distance $\hat{x}$ can be assigned to $S(x)$ (see, e.g., [4]); the connection is $\hat{x} \cdot \hat{f} = 1/2$.

For realistic problems the variable x can represent a position or time coordinate, a wavelength or another physical quantity. The concept of finite sampling point distance has been successfully utilized also in many other physical and technical problems, this rests on the fact that frequently the relation between the output (or action) quantity and the input (or cause) quantity has the form of equation (2.1), which must be regarded as the linear term of a Volterra expansion of the functional $y_{\text{out}}[y_{\text{in}}(x)]$. In information and communication technics time-dependent signal functions are recorded at the output of physically realizable filters. Since always a finite total signal duration τ must be assumed, the information content of the, in reality, continuously running signal is given by a finite number $m = \tau/\hat{x} + 1$ of signal ordinates $y(x_0), \dots, y(x_m)$; this statement is in agreement with the sampling point theorem [4]. In the stationary and transient spectrometry the concept of finite sampling point distance for spectra leads to general, experimentally verified formulas for the optimum conditions under which the input can be determined from the measured output [5, 6]. Also basic physical problems can be advantageously treated: The quantum-state functions of the position and momentum representation are Fourier transforms due to the basic commutation relation [7]; since their moduli squared have the character of probability densities, they are square-integrable and the sampling point concept is applicable.

Let us now derive the explicit form of $S(x)$ from the abovedefined PI-properties. Assuming $x_n = n\hat{x}$, the properties (2.3) immediately lead to

$$S(x) = \sum_{n=0}^{m} y(n\hat{x})\, s_n(x) \tag{2.4}$$

with $s_n(x) = s(\pi x/\hat{x} - n\pi)$ and $s(a) = \sum_{j=0}^{\infty} (-1)^j a^{2j} (2j+1)!$. For $a \neq 0$ there holds $s(a) = a^{-1} \cdot \sin a$. Thus, due to $s_n(n\hat{x}) = 1$ and $s_n(n'\hat{x}) = 0$ for $n' \neq n$, the properties of the functions $s_n(x)$ guarantee $S(n\hat{x}) = y(n\hat{x})$ and thus, the interpolation problem is uniquely solved. The Fourier transforms $\mathbf{s}_n(f)$ of the expansion functions vanish for $\hat{f} > 1/2\hat{x}$; thus, the function $S(x)$ has the required finite limiting frequency $\hat{f} = 1/2\hat{x}$ and the finite sampling point distance $\hat{x}$. Without loss of generality we may assume, that x is a dimensionless quantity and $\hat{x}$ equals 1. Then there follows the PI-interpolation function

$$S(x) = \sum_{n=0}^{m} y(n)\, s_n(x)\,,$$

where

$$s_n(x) = s(\pi x - \pi n) = \begin{cases} 1 & \text{for } x - n = 0 \\ \dfrac{\sin(\pi x - \pi n)}{\pi x - \pi n} & \text{for } x - n \neq 0\,. \end{cases} \tag{2.5}$$

So far we have studied the physical-technical reasons for the form (2.5) of the function $S(x)$. In the following we will discuss its mathematical properties and compare them with those of the Lagrange and spline functions. If the ordinate values $y(n)$ (which may now be regarded as ordinates at the points of support) are given, the interpolation problem in the PI-case as well as in the Lagrange-case is solved uniquely; this does not hold in the spline-case as can be concluded from equation (1.3). Let us now consider the influence of the interval length m. In the PI-case the expansion functions $s_n(x)$ are independent of m. In the Lagrange-case there exists a strong alteration, the degree of the polynomials $l_{n,m}(x)$ increases with growing m; this fact leads, at least near the edges of the interval, to a drastic effect concerning the derivative, as it will be quantitatively seen in Section 3. However, one can hardly assume that there exist realistic conditions, for which the contribution of one additional ordinate value at one edge of the interval considerably strengthens the derivative of the function at the other edge. This undesirable fact is avoided in the PI-case. In the spline-case the polynomials $c_{n,m}(x)$ also change with changing m; but there is only a slight influence.

With the help of the expressions in equations (2.5), (1.1), (1.2) we will estimate derivatives and integrals over $[0, m]$. The results depend on the interpolation type chosen and on the set $\{y(n)\}$. The latter means that for a quantitative estimation sufficiently detailed information about the ordinate ensemble is needed. On the other hand we aim at results to be applied in a wide range of validity. To meet both these requirements we have to consider, so to say, a large number of various ordinate sets $\{y(n)\}$. This can practically be done by introducing a certain probability distribution for the $y(n)$. Moreover, we choose the following general conditions: The $y(n)$ are independent of one another and possess the ensemble mean values

$$\langle y(n) \rangle = 0\,, \qquad \langle y(n)\, y(n' \neq n) \rangle = 0\,, \qquad \langle y^2(n) \rangle = q^2\,. \tag{2.6}$$

It should be mentioned that these assumptions include the representative case of a Gaussian signal. Equation (2.6) leads to

$$\langle S(x) \rangle = 0\,, \qquad \langle L(x) \rangle = 0\,, \qquad \langle C(x) \rangle = 0\,. \tag{2.7}$$

3. Derivatives and the strength of interpolation vibrations

First we will consider the differentiability in the interval $[0, m]$. Equation (2.5) shows that the derivatives $S^{(k)}(x)$ are defined for all k; the value of $S^{(k)}(x)$ only depends on the $y(n)$-values. All the derivatives $L^{(k)}(x)$ are defined, but in contrast to $S(x)$, the derivatives $L^{(k>m)}(x)$ vanish independently of the $y(n)$. The spline function $C(x)$ gives up the continuity of the third derivative. This qualitative consideration exhibits a superiority of the PI-function $S(x)$. In the following we will discuss further quantitative advantages of $S(x)$ by means of ensemble mean values with respect to the $y(n)$.

From equations (2.5) and (2.6) one directly obtains the *variance* V and the *reduced variance* v of the first derivative of $S(x)$ at x, namely

$$V(S'; m; x) = \langle |S'(x) - \langle S'(x) \rangle|^2 \rangle = \sum_{n=0}^{m} q^2 |s_n'(x)|^2 = q^2 v(S'; m; x)\,. \tag{3.1}$$

V is proportional to the (ensemble) mean square value of the ordinates $y(n)$. Since we are not interested in the absolute value of the variance, we have introduced the reduced variance v by eliminating q^2. In general, derivatives linearly depend on the values of the function; therefore we will mainly discuss the root mean square (r.m.s.) quantity $w = v^{1/2}$. Since $\langle S'(x) \rangle = 0$, the quantity $w(S'; m; x)$ can be regarded as a measure for the fluctuations of the function $S(x)$, that means the strength of interpolation vibrations.

A deeper insight can be gained by introducing mean values that are related to the average over the whole interval $[0, m]$. To display the situation we distinguish between quantities that only are ensemble-averaged at a fixed point x, like $w(S'; m; x)$, and interval- and ensemble-averaged quantities, like

$$w(S'; m) = \left[\sum_{n=0}^{m} \frac{1}{m} \int_0^m dx |s_n'(x)|^2 \right]^{1/2}. \tag{3.2}$$

The reduced interval-averaged variance possesses an upper bound, since the properties of $s_n(x)$ yield

$$v(S'; m) = \frac{2}{m} \sum_{n=1}^{m} \int_0^n dx\, |s_n'(x)|^2 < 2 \cdot \int_0^\infty dx |s_0'(x)|^2\,; \tag{3.3}$$

thus, the corresponding r.m.s. quantity $w(S'; m)$ also possesses an upper bound $w_{\text{ub}}(S'; m)$, which can be numerically given by 1.82. For $m = 3$ and $m = 4$ there results $w(S'; 3) \approx 1.6$ and $w(S'; 4) \approx 1.7$; the usual rounding convention is used (that means, e.g., $1.55 \leq w(S'; 3) < 1.65$). Moreover, we will discuss values at a fixed point, namely at the edge of the interval $x = 0$. There results from (3.1)

$$v(S'; m; 0) = \sum_{n=1}^{m} \frac{1}{n^2} \quad \text{and} \quad v(S'; m \to \infty; 0) \to \frac{\pi^2}{6}\,. \tag{3.4}$$

For the corresponding r.m.s. quantities there follows $w(S'; 4; 0) \approx 1.2$ and the upper bound $w_{\text{ub}}(S'; m; 0) = 1.29$.

Combining the above results we arrive at

$$w(S'; 4; 0) \approx 1.2 < w_{ub}(S'; m; 0) = 1.29 < w(S'; 4) \approx$$
$$\approx 1.7 < w_{ub}(S'; m) = 1.82; \qquad (3.5)$$

$w(S'; m)$ increases with growing m, but it does not exceed the value 1.82. The interval-averaged r.m.s. fluctuations $w(S'; m)$ surpasses the fluctuations assigned to the interval edge $x = 0$, since $s_0'(x) = 0$.

Let us now study the behavior of the Lagrange function. At $x = 0$ there results the reduced variance

$$v(L'; m; 0) = \sum_{n=0}^{m} |l'_{n,m}(0)|^2 \qquad (3.6)$$

for the first derivative. The corresponding r.m.s. values are $w(L'; 1; 0) \approx 1.4$; $w(L'; 2; 0) \approx 2.5$; $w(L'; 3; 0) \approx 3.8$; $w(L'; 4; 0) \approx 5.6$. It can be concluded [2] from the defining equation (1.1) that $w(L'; m; 0)$ strongly increases with increasing m. Numerical interval-averaged values are $w(L'; 3) \approx 1.8$; $w(L'; 4) \approx 2.1$. Obviously the strength of L' at the interval edges exceeds the interval-averaged value, since in the interior of the interval the polynomial $L(x)$ exhibits a smoother behavior than near the interval edges.

From (1.2) and (1.3) one obtains by a straightforward calculation the reduced variance of the first derivative of the spline function

$$v(C'; m) = 2 +$$
$$+ \frac{1}{m} \sum_{n=0}^{m-1} \frac{4\langle|C''(n)|^2\rangle + 4\langle|C''(n+1)|^2\rangle + 7\langle C''(n)\, C''(n+1)\rangle}{180q^2}. \qquad (3.7)$$

Numerical evaluation yields $w(C', 4) \approx 1.6$ and a slight increase with increasing m. The three interpolation procedures can be compared with help of the inequality

$$w(C'; 4) \approx 1.6 < w(S'; 4) \approx 1.7 < w_{ub}(S'; m) =$$
$$= 1.82 < w(L'; 4) \approx 2.1\,. \qquad (3.8)$$

This relation reveals that the quantity assigned to the spline procedure with respect to the avoidance of interpolation vibrations only slightly (some percent) exceeds the quantity derived in the PI-case. This holds, although, expressively in order to get the smoothest function in $C^2(0, m)$, a remarkable loss of generality has been allowed for the spline function with respect to highest order of continuous differentiability.

This favourable situation for the function $S(x)$ with PI-properties also remains unchanged, if another sensible measure for the strength of interpolation vibrations is used, namely the second derivative $C''(x)$ as already introduced in the total curvature equation (1.4). There exists, as for the first derivative, an upper bound for the interval-averaged r.m.s. value $w(S''; m)$ of the second derivative, namely

$$w(S''; m) < [2 \int_0^\infty dx |s_0''(x)|^2]^{1/2} < 4.5 = w_{ub}(S''; m) \qquad (3.9)$$

and this upper bound relatively slightly exceeds the corresponding spline quantities $w(C''; m)$.

4. Integrals and summation error

From the defining relation (2.5) for $S(x)$ there follows the integral

$$\int_0^m dx\, S(x) = Q(S, m) - A(S, m)\,, \qquad (4.1)$$

where

$$Q(S, m) = \tfrac{1}{2}\,[S(0) + S(m)] + \sum_{n=1}^{m-1} S(n) =$$
$$= \tfrac{1}{2}\,[y(0) + y(m)] + \sum_{n=1}^{m-1} y(n)\,.$$

$Q(S, m)$ may be called the *normalized sum of the ordinate values*, since it is chosen in such a way that the interval length of the corresponding step function equals that of the integral. $A(S, m)$ may be regarded as the error characterizing the replacement of integration by summation. This *summation error* turns out to be

$$A(S, m) = \sum_{n=0}^{m} y(n)\, D(S; n; m)\,, \qquad (4.2)$$

where

$$D(S; n; m) = \begin{cases} T[m\pi] & \text{for } n = 0, m \\ T[n\pi] + T[(m-n)\pi] & \text{for } n = 1, \ldots, m-1\,. \end{cases}$$

$T[n\pi]$ is connected with the integral sine and tends to the given asymptotic expression for large n:

$$T[n\pi] \equiv$$
$$\equiv \frac{1}{2} - \frac{1}{\pi}\, \mathrm{Si}[n\pi] \cong (-1)^n\, n^{-1}\pi^{-2}[1 - n^{-2}\pi^{-2}2! + n^{-4}\pi^{-4}4! \ldots]\,. \qquad (4.3)$$

To get relatively simple relations in the Lagrange case we take m to be odd. Then we have the summation error

$$A(L, m) = \sum_{n=0}^{m} y(n)\, D(L; n, m) \qquad (4.4)$$

where $D(L; n; m) = \sum_{n=0}^{(m-1)/2} \frac{B_{2j}}{(2j)!}\, [l_{n,m}^{(2j-1)}(m) - l_{n,m}^{(2j-1)}(0)]$ with Bernoulli numbers B_{2j}.

Equations (4.2) and (4.4) show that the integral value of the PI-function $S(x)$ and the Lagrange function are uniquely determined by the ordinate values $y(0), \ldots, y(m)$; this also holds for the spline function. The summation errors $A(S, m)$ and $A(L, m)$ show qualitative differences: in the PI-case the coefficients $D(S; n; m)$ are given by the values of the integral sine $\mathrm{Si}[n\pi]$, while in the Lagrange case the $D(L; n; m)$ are rational numbers. Now we will consider quantitative differences. The r.m.s. values of the reduced summation error are

$$\left.\begin{aligned} [\langle A^2(S, m)\rangle]^{1/2}/q &= [\sum_{n=0}^{m} D^2(S; n; m)]^{1/2}\,; \\ [\langle A^2(L, m)\rangle]^{1/2}/q &= [\sum_{n=0}^{m} D^2(L; n; m)]^{1/2}\,. \end{aligned}\right\} \qquad (4.5)$$

There result, already for small m, relatively large differences; for instance in the case $m = 3$, the r.m.s. reduced summation error is ≈ 0.073 for $S(x)$ and ≈ 0.25 for $L(x)$. This example reveals that also with respect to numerical integration a certain superiority of the PI-interpolation exists.

Finally we will show that the summation error $A(S, m)$ in the PI-case can be approximated by an expression fitted to practical problems. To get symmetrical terms and a comparatively better estimation we take m to be an odd number. As a rule the ordinates $y(n)$ cannot be given or measured exactly; let the mean error be $|\Delta y|$. Furthermore, let us suppose that the ordinates $y(n)$ have the same order of magnitude q. Under these conditions it is advantageous to subdivide the right-hand side of (4.2) into "great" and "small" terms according to

$$A(S, m) = \sum_{n=1}^{n_0} T[n\pi] \cdot [y(n) + y(m-n)] + \sum_{n} \text{small terms}\,, \qquad (4.6)$$

where n_0 is a suitable integer smaller than $(m-1)/2$. The second sum on the right-hand side may contain all terms for which the coefficients T are much less than $|\Delta y|/q$, so that the second sum can be neglected. Thus, one has the *approximate summation error*:

$$A_{app}(S, m) = -0.0895 \cdot [y(1) + y(m-1)] +$$
$$+ 0.0486 \cdot [y(2) + y(m-2)] + \ldots + T[n_0\pi] \times$$
$$\times [y(n_0) + y(m-n_0)]\,. \qquad (4.7)$$

The usefulness of this approximating formula rests on the fact that $T[n\pi]$ decreases asymptotically with $1/n\pi^2$ for large n. Equation (4.7) exhibits that under suitable conditions only the ordinates near the interval edges yield significant contributions to the summation error of the PI-function $S(x)$. To some extent an analogous feature reveals the Taylor-expanded function $E(x)$; its summation error can be given with the help of the Euler-Maclaurin formula:

$$A(E, m) = \sum_{j=1}^{(k-1)/2} \frac{B_{2j}}{(2j)!}\, [E^{(2j-1)}(m) - E^{(2j-1)}(0)] +$$
$$+ p_k(\vartheta m)\, E^{(k)}(\vartheta m)\,, \qquad (4.8)$$

where

$0 < \vartheta < 1\,.$

$p_k(x)$ is a polynomial of k-th degree. Here, additionally to the ordinates $E(0), \ldots, E(n), \ldots, E(m)$ the derivatives $E^{(1)}, E^{(3)}, \ldots$

at the interval edges are needed to determine the integral $\int_0^m dx\, E(x)$ via $Q(E, m)$ and $A(E, m)$. In the PI-case the knowledge on the ordinates near the interval edges approximately suffices.

5. Summary

It has been shown that objective reasons in physics and technics frequently suggest the introduction of an interpolation function $S(x)$ with a finite limiting frequency and a finite sampling point distance, respectively. This condition and the knowledge of a given set $\{y(n)\}$ of ordinate values suffice to construct uniquely $S(x)$, which we may call *physically implied interpolation function.*

Independently of its physically implied introduction, the mathematical properties of the function $S(x)$ have been studied. Subjecting the ordinate values to a plausible probability distribution, sensible measures for the strength of interpolation vibrations and the integration remainder term could be evaluated that are applicable in a wide range of validity. The qualitative and quantitative comparison of the physically implied interpolation function $S(x)$ with the Lagrange and spline interpolation functions shows a certain superiority of $S(x)$; it has either the same quality under more general assumptions or a better quality under the same assumptions.

I gratefully acknowledge valuable discussions with Prof. Dr. H. J. Glaeske, Dr. R. Kühmstedt, Dr. H. Toparkus, Prof. Dr. W. Wallisch.

References

1 Schubert, M.: Forschungsergebnisse d. Univ. Jena N14 (05/1987).
2 Schubert, M.: Forschungsergebnisse d. Univ. Jena N17 (07/1987).
3 Stoer, J.: Einführung in die Numerische Mathematik I, Heidelberg 1972, Kap. 2.4.
4 Goldman, S.: Information theory, New York 1955, Sect. 2.1.
5 Schubert, M.: Exp. Techn. d. Phys. 6 (1958), 203; 14 (1966), 1.
6 Gase, R.; Schubert, M.: Optica Acta 29 (1982), 1331.
7 Schubert, M.; Weber, G.: Quantentheorie — Grundlagen, Methoden, Anwendungen, Berlin 1983, Kap. 8.4.2.

Received July 20, 1987, in revised form November 18, 1987

Address: Professor Dr. Max Schubert, Department of Physics, Friedrich-Schiller-University Jena, Max-Wien-Platz 1, Jena, DDR-6900, G.D.R.

ZAMM · Z. angew. Math. Mech. 69 (1989) 2, 109–111

Reitgruber, W.; Tichy, R. F.

Numerische Lösung gewisser partieller Differentialgleichungen mittels guter Gitterpunkte

In [6] wurde mit der auf N. M. Korbov [4] und E. Hlawka [2] zurückgehenden Methode der guten Gitterpunkte das Cauchysche Anfangswertproblem für eine spezielle Klasse partieller Differentialgleichungen näherungsweise gelöst. Insbesondere kann damit die Wellengleichung (vgl. auch [3]) und die Plattengleichung behandelt werden. Im folgenden wird eine weitreichende Verallgemeinerung und Quantifizierung der Fehlerabschätzungen skizziert. Sodann werden die Ergebnisse numerischer Tests diskutiert. Es stellt sich heraus, daß dieses Verfahren für den dreidimensionalen Fall sehr vorteilhaft werden kann, für höhere Dimensionen erkennt man aber nachteilige Effekte.

Es bezeichne $E_s^\alpha(C)$, $\alpha > 1$, $C > 0$, den Raum aller periodischen Funktionen

$$f(x) = \sum_{h \in Z^s} c(h) \exp(2\pi i\langle h, x\rangle); \qquad x = (x_1, \dots, x_s),$$

welche die Abschätzung $c(h) \leqq C/||h||^\alpha$ erfüllen; dabei ist $h = (h_1, \dots, h_s)$, $||h|| = \prod_{j=1}^{s} \max\{1, |h_j|\}$ und $\langle h, x\rangle = \sum_{j=1}^{s} h_j x_j$.

Wir behandeln das Anfangswertproblem

$$D_t u(t, x) = -(-D)^p u(t, x), \qquad \left.\frac{\partial^j}{\partial t^j} u(t, x)\right|_{t=0} = f_j(x), \qquad j = 0, \dots, k-1, \tag{1}$$

wobei p, k natürliche Zahlen und D_t, D die Differentialoperatoren

$$D_t := \sum_{j=0}^{k} b_j \frac{\partial^j}{\partial t^j}, \qquad D := \sum_{i=1}^{s} \sum_{j=1}^{s} a_{ij} \frac{\partial^2}{\partial x_i\, \partial x_j}$$

sind; $A = ||a_{ij}||$ bezeichnet eine positiv definite Matrix, und f_j sind gegebene Funktionen aus $E_s^\alpha(C)$, $\alpha - 2p > 1$.

Um (1) zu lösen, benützt man die Fourierreihenentwicklung

$$u(t, x) = \sum_{h \in Z^s} c(t, h) \exp(2\pi i\langle h, x\rangle)$$

der Lösung und erhält

$$D_t u(t, x) = \sum_{h \in Z^s} D_t c(t, h) \exp(2\pi i\langle h, x\rangle), \tag{2}$$

$$-Du(t, x) = \sum_{h \in Z^s} c(t, h)\, 4\pi^2[h, h] \exp(2\pi i\langle h, x\rangle), \tag{3}$$

wobei das Skalarprodukt $[h, h]$ durch hAh^T gegeben ist. Setzt man diese Ergebnisse in die zu lösende Differentialgleichung ein, folgt die Beziehung

$$\sum_{h \in Z^s} D_t c(t, h) \exp(2\pi i\langle h, x\rangle = $$

$$= -\sum_{h \in Z^s} c(t, h)\, (4\pi^2[h, h])^p \exp(2\pi i\langle h, x\rangle), \tag{4}$$

woraus mittels Koeffizientenvergleiches die Differentialgleichung

$$D_t c(t, h) = -c(t, h)\, (4\pi^2[h, h])^p \tag{5}$$

für die Fourierkoeffizienten der Lösung erhalten wird. Die Anfangswerte ergeben sich aus der Darstellung der Funktionen f_j als Fourierreihen durch die Beziehung

$$\sum_{h \in Z^s} F_j(h) \exp(2\pi i\langle h, x\rangle) = f_j(x) = \left.\frac{\partial^j}{\partial t^j} u(t, x)\right|_{t=0} = $$

$$= \sum_{h \in Z^s} \left.\frac{\partial^j}{\partial t^j} c(t, h)\right|_{t=0} \exp(2\pi i\langle h, x\rangle),$$

woraus über Koeffizientenvergleich

$$F_j(h) = \left.\frac{\partial^j}{\partial t^j} c(t, h)\right|_{t=0} \tag{6}$$

die gesuchten Anfangswerte gegeben sind.

Um zu einer expliziten Darstellung der Lösung zu gelangen, beschränkt man sich auf den Fall $D_t = \partial^k/\partial t^k$. Damit besitzt die Differentialgleichung für die Fourierkoeffizienten die Lösung

$$c(t, h) = \sum_{j=1}^{k} C_j(h) \exp\left(t\eta_k^j \sqrt[k]{-(4\pi^2[h, h])^p}\right) \tag{7}$$

wobei η_k^j die k-ten Einheitswurzeln bezeichnen und die Konstanten $C_j(h)$ durch die Anfangsbedingungen über das lineare Gleichungssystem

$$F_j(h) = \left.\frac{\partial^j}{\partial t^j} c(t, h)\right|_{t=0} = \left.\frac{\partial^j}{\partial t^j} \sum_{j_1=1}^{k} C_{j_1}(h) \exp\left(t\eta_k^{j_1} \sqrt[k]{-(4\pi^2[h, h])^p}\right)\right|_{t=0}$$

$$= \sum_{j_1=1}^{k} C_{j_1}(h) \left(\eta_k^{j_1} \sqrt[k]{-(4\pi^2[h, h])^p}\right)^j, \qquad j = 0, \dots, k-1, \tag{8}$$

bestimmt werden können. Man kann allerdings nur dann sinnvolle Lösungen erwarten, wenn die Koeffizienten $C_{j_1}(h)$, bei denen $\eta_k^{j_1} \sqrt[k]{-(4\pi^2)^p}$ einen positiven Realteil besitzt, verschwinden. Andernfalls bekommt man für $t > 0$ Fourierreihen für $u(t, .)$, deren Koeffizienten exponentiell in $[h, h]$ wachsen und deshalb nicht konvergieren können. Im Fall $k = 2$ erübrigt sich diese Bedingung, da die Größen $\pm i(2\pi)^p$ rein imaginär sind und deshalb keinen positiven Realteil besitzen können.

Um die Größe der $C_{j_1}(h)$ in den Griff zu bekommen, kann man zur Darstellung der Lösung die Cramersche Regel verwenden. Kurze Rechnung ergibt für die Systemdeterminante den Wert $\mathrm{Det} = A[h, h]^{p^2(p-1)/2k}$ mit einer Konstanten A und für die

Determinante Det_{j_1} der Matrix, die aus der Systemmatrix durch Vertauschen der j_1-ten Spalte mit dem Vektor $(F_1(h), \ldots, F_k(h))^T$ entsteht, die Abschätzung

$$|\mathrm{Det}_{j_1}| \leqq A_{j_1} \frac{C}{||h||^\alpha} [h, h]^{p(p-1)(p-2)/2k} .$$

Somit gilt die Ungleichung

$$|C_{j_1}(h)| \leqq \frac{B}{||h||^\alpha [h, h]^{p(p-1)/k}} \tag{9}$$

für eine positive Konstante B, falls die Systemdeterminante ungleich Null ist. Eine ähnliche Ungleichung gilt natürlich auch für die Fourierkoeffizienten $c(t, h)$, falls die C_{j_1} verschwinden, die einen exponentiell steigenden Anteil beitragen. Dann gilt

$$|c(t, h)| \leqq B'/||h||^\alpha [h, h]^{p(p-1)/k} . \tag{10}$$

Da $||h||^2 \geqq B''[h, h]$ mit einer Konstanten $B'' > 0$ gilt, kann weiters die absolute Konvergenz der Fourierreihe, die durch das Einsetzen in die Differentialgleichung (3) entstanden ist, aus

$$|c(t, h) (4\pi^2[h, h])^p)| \leqq B'''/||h||^{(\alpha-2p)} \tag{11}$$

und der Bedingung $\alpha - 2p > 1$ gefolgert werden. Somit läßt sich der verwendete Ansatz im nachhinein rechtfertigen.

Kurz zusammengefaßt ist es zur numerischen Lösung notwendig, die Fourierkoeffizienten $F_j(h)$ der Funktionen $f_j(h)$ genau zu bestimmen.

Die *Methode der guten Gitterpunkte* zur Lösung des Problems (1) besteht nun darin, mit Hilfe des folgenden Lemmas die Fourierkoeffizienten zu approximieren. Dieses Lemma ist eine quantifizierte Fassung des in [6] formulierten Resultats; der Beweis wird induktiv geführt und findet sich ausführlich in [5].

Lemma: *Es sei $\alpha > 1$ und $f \in E_s^\alpha(C)$. Ferner sei $g \in Z^s$ ein ganzzahliger Gitterpunkt, der für einen gegebenen Modul m die Ungleichung*

$$M := \min_{\langle h, g\rangle \equiv 0 (\mathrm{mod}\ m) h \neq 0} ||h|| - 1 > 2^{s(2\alpha-1)(\alpha-1)} \tag{12}$$

erfüllt. Dann gilt die Fehlerabschätzung ([.] *bezeichnet den Ganzteil, $\omega_n = ng/m$*)

$$\sup_{f \in E_s^\alpha(C)} \sum_{||h|| < [M^{\alpha/(2\alpha-1)}]} \left| F(h) - \frac{1}{m} \sum_{n=1}^{m} f(\omega_n) \exp(-2\pi i \langle \omega_n, h\rangle) \right| \leqq \gamma M^{-[\alpha(\alpha-1)]/(2\alpha-1)]+1} , \tag{13}$$

wobei $F(h)$ die Fourierkoeffizienten von f und γ die Konstante

$$\gamma = \frac{C\big(2^{(\alpha+1)}3(2\zeta(\alpha) + 1)\big)^s s!^2}{(\alpha - 1)}$$

bezeichnen (ζ ist die Riemannsche ζ-Funktion).

Mit diesem Hilfssatz kann dann wie in [6] das allgemeinere Anfangswertproblem näherungsweise gelöst werden. Wir gehen im folgenden nur auf den wichtigen Spezialfall $k = 2$, $F_1(0) = 0$ ein. In diesem Fall ergibt sich aus den obigen allgemeinen Betrachtungen eine einfache Reihendarstellung der Lösung von (1), aus der durch Approximation der Fourierkoeffizienten die Näherungslösung

$$\tilde{u}(t, x) := \tilde{F}_0(0) + \sum_{0 \neq ||h|| < [M^{\alpha/(2\alpha-1)}]} \frac{1}{2} \left(\left(\tilde{F}_0(h) + i \frac{\tilde{F}_1(h)}{(2\pi \sqrt{[h, h]})^p} \right) \times \exp[-it(2\pi \sqrt{[h, h]})^p] + \left(\tilde{F}_0(h) - i \frac{\tilde{F}_1(h)}{(2\pi \sqrt{[h, h]})^p} \right) \times \exp[it(2\pi \sqrt{[h, h]})^p] \right) \exp(2\pi i \langle h, x\rangle) \tag{14}$$

mit den Koeffizienten

$$\tilde{F}_j(h) := \frac{1}{m} \sum_{n=1}^{m} f(\omega_n) \exp[-2\pi i \langle \omega_n, h\rangle] \tag{15}$$

gewonnen werden kann. Das obige Lemma liefert die explizite Fehlerabschätzung

$$|u(t, x) - \tilde{u}(t, x)| \leqq M^{-[\alpha(\alpha-1)]/(2\alpha-1)]} \left(\gamma M + \frac{C3^s s!}{\alpha - 1} \right) \left(1 + \left(2\pi \min_{h \neq 0} \sqrt{[h, h]} \right)^{-p} \right). \tag{16}$$

Numerische Tests

Als Testfunktion wird

$$H_6(x) = \prod_{j=1}^{s} \left(\frac{(2\pi)^6}{6!} B_6(\{x_j\}) + 1 \right) - 1$$

verwendet; B_6 bezeichnet das 6. Bernoullipolynom und $\{x\}$ den Nachkommateil von x. Es gilt $H_6 \in E_s^6(1)$. In Tabelle 1 sind die Ergebnisse bezüglich der Problemstellung $p = 2$ (Plattengleichung), $f_0(x) = f_1(x) = H_6(x)$ mit nichttrivialen Matrizen A, aufgelistet. Dabei findet sich in der 1. Spalte die Dimension s in der 2. Spalte der Modul m des verwendeten Gitterpunktes, in der 3. Spalte stehen die Koordinaten des guten Gitterpunktes g und in der 4. Spalte die Güte M dieses Gitterpunktes. Die 5. Spalte enthält den Gesamtfehler, der bei der Lösung der partiellen Differentialgleichung über diesen Gitterpunkt entstanden ist, und die 6. Spalte enthält den Gesamtfehler, der durch die Approximation der Integrale über ein äquidistantes Gitter entstanden ist.

	m	g	M	F_1	F_2
2	89	(1, 55)	33	$0{,}63 \cdot 10^{-3}$	$1{,}5 \cdot 10^{-3}$
2	144	(1, 89)	54	$0{,}68 \cdot 10^{-3}$	$0{,}53 \cdot 10^{-3}$
2	233	(1, 144)	88	$1{,}77 \cdot 10^{-3}$	$0{,}48 \cdot 10^{-3}$
3	88	(1, 18, 28)	9	2,34	0,18
3	185	(1, 26, 64)	19	0,13	0,11
3	366	(1, 50, 128)	35	$6{,}8 \cdot 10^{-3}$	0,09
5	772	(1, 154, 170, 230, 256)	9	139,1	29,1

In der 2. Dimension können also noch keine signifikanten Unterschiede zwischen den beiden Methoden festgestellt werden, wogegen in der 3. Dimension bereits die deutlich bessere Konvergenzordnung der Methode der guten Gitterpunkte zum Ausdruck kommt. In der 5. Dimension wurde zwar der beste bekannte Gitterpunkt verwendet, für gute Ergebnisse wären aber noch weitaus bessere Gitterpunkte nötig, die aber nur unter großem Aufwand gefunden werden können.

Aus den Detailergebnissen [5] kann außerdem folgender Effekt herausgelesen werden: Bei guten Gitterpunkten mit relativ kleinem M, das ein Gütemaß für diesen Punkt darstellt, kommt es bei der Berechnung der Fourierkoeffizienten zu Interferenzerscheinungen zwischen der Struktur der Stützpunkte und der zu integrierenden Funktion, wodurch einige wenige Fourierkoeffizienten sehr schlecht approximiert werden und Gesamtfehler stark vergrößern.

Diskussion

1. Die theoretischen Fehlerabschätzungen sind sehr pessimistisch, wodurch die tatsächlichen Fehler weit unter den erwarteten Fehlern liegen.

2. In der Dimension 2 sind die Ergebnisse mit der Methode der guten Gitterpunkte etwa gleich gut wie bei der Wahl eines äquidistanten Gitters.

3. Ist die Dimension 3, liefert die Methode der guten Gitterpunkte deutlich bessere Ergebnisse. Allerdings kann bei der Wahl eines zu kleinen Gitterpunktes bereits das Interferenzphänomen auftreten.

4. In höheren Dimensionen tritt mit den verfügbaren Gitterpunkten das Interferenzphänomen stark auf, wodurch keine guten Ergebnisse erhalten werden. Bessere Gitterpunkte sind aber in diesen Dimensionen nur sehr schwer zu bestimmen.

Schlußbemerkung

In [7] werden praktische Gitterpunkte statt der guten Gitterpunkte zur numerischen Integration empfohlen, die zwar leicht berechenbar, jedoch nicht optimal sind. Diese wurden in [5] ebenfalls getestet, es ergab sich aber ein verstärktes Auftreten des Interferenzphänomens (entsprechend der deutlich kleineren Güte dieser Gitterpunkte). Das Programm, mit dem die Berechnungen durchgeführt wurden, kann auch in [5] nachgelesen werden.

Literatur

1 BOURDEAU, M.; PITRE, A.: Tables of good lattices in four and five dimensions. Numer Math. 47 (1985), 39–43.
2 HLAWKA, E.: Zur angenäherten Berechnung mehrfacher Integrale. Monatsh. Math. 66 (1962), 140–151.
3 HUA, L. K., WANG, Y. Applications of number theory to numerical analysis. Springer-Verlag Berlin etc. 1981.
4 KOROBOV, N M. The approximate computation of multiple integrals. Doklad. Akad. Nauk. SSSR 124 (1959), 1207–1210.
5 REITGRUBER, W. Numerische Analysis und Funktionen von beschränkter Schwankung. Diss., techn.-naturwiss. Fak., TU Wien 1987.
6 TICHY, R. F.: Ein Approximationsverfahren zur Lösung partieller Differentialgleichungen. Z. angew. Math. Mech. 68 (1988) 3, 187–189.
7 ZINTERHOF, P.: Gratis lattice points for multidimensional integration. Computing 38 (1987), 347–353.

Eingegangen am 1. Februar 1988

Anschrift: Dr. WOLFGANG REITGRUBER, Dr. ROBERT F. TICHY, Institut für Analysis, Technische Mathematik und Versicherungsmathematik, Wiedner Hauptstraße 8–10, A-1040 Wien, Österreich

ZAMM · Z. angew. Math. Mech. 69 (1989) 2, 111–112

WOLFERSDORF, L. v.

Stokes Flow Past a Sphere with Permeable Surface

Introduction

The Stokes solution for the slow motion past an impermeable sphere in a viscous liquid is well-known (cf., for instance, [1], § 35). Also the slow motion of a viscous fluid past a permeable sphere in a uniform stream can be calculated in analogous analytic form if the fluid in the permeable sphere obeys the Darcy law [2]. In this note we give the solution of the corresponding Stokes flow past a sphere with permeable surface and also past a sphere with permeable surface containing a concentric impermeable sphere in its interior. Accordingly to [2], as boundary conditions at the permeable surface of the sphere we choose a conjugacy condition of Darcy type and the no-slip condition for the tangential components of the velocity.

The case of a single sphere

We consider a stationary sphere of radius a immersed in a uniformly streaming viscous fluid without body forces. The sphere has a permeable surface and the flow outside and inside the sphere satisfies the stationary Stokes equations

$$\operatorname{grad} p = \mu \Delta \boldsymbol{v}\,, \qquad \operatorname{div} \boldsymbol{v} = 0$$

leading to the scalar system

$$\partial p/\partial x = \mu \Delta v_x\,, \qquad \partial p/\partial y = \mu \Delta v_y\,, \qquad \partial p/\partial z = \mu \Delta v_z\,, \tag{1}$$

$$\Delta p = 0 \quad (\Delta \text{ Laplace operator})\,,$$

where p is the (hydrodynamic) pressure, μ is the dynamic viscosity, and v_x, v_y, v_z are the Cartesian components of the velocity $\boldsymbol{v}$. In the sequel we use spherical coordinates (r, φ, θ), where θ is the angle composed by the radius vector with the z-axis which is taken as the direction of the free stream. Further, let be $\varrho = \sqrt{x^2 + y^2} = r \sin \theta$. The flow is axially symmetric, i.e., independent of φ.

Additionally to the system (1) we have the following boundary conditions, where + and − refer to the inner and outer boundary values, respectively.

1. The normal component v_r of velocity is continuous at the boundary of the sphere and apart from the sign equal to the filtration velocity V through the boundary:

$$v_r^+(a, \theta) = v_r^-(a, \theta) = -V(\theta)\,. \tag{2}$$

2. The filtration velocity V is proportional to the difference of the hydrodynamic pressure $p^- - p^+$:

$$V(\theta) = \frac{k}{\mu} \left(p^-(a, \theta) - p^+(a, \theta)\right) \tag{3}$$

with a permeability coefficient $k > 0$. (The friction pressure p_{rr} in normal direction will be continuous across the boundary of the sphere.)

3. The fluid does not slip relative to the sphere:

$$v_\theta^+(a, \theta) = v_\theta^-(a, \theta) = 0\,. \tag{4}$$

4. The velocity at infinity is that of the free stream q:

$$v_r(r, \theta) \to q \cos \theta\,, \qquad v_\theta(r, \theta) \to -q \sin \theta \quad \text{as} \quad r \to \infty\,. \tag{5}$$

Further, without loss of generality, we suppose $p = 0$ at infinity.

The differential system (1) is solved in usual way introducing a stream function Ψ by means of

$$v_r = -(1/\varrho)\, \partial\Psi/\partial z\,, \qquad v_z = (1/\varrho)\, \partial\Psi/\partial\varrho\,. \tag{6}$$

Then for Ψ there holds the equation

$$\bar{\Delta}\bar{\Delta}\Psi = 0 \tag{7}$$

with the Stokes operator

$$\bar{\Delta} := \frac{\partial^2}{\partial z^2} + \frac{\partial^2}{\partial \varrho^2} - \frac{1}{\varrho}\frac{\partial}{\partial \varrho} = \frac{\partial^2}{\partial r^2} + \frac{\sin\theta}{r^2}\frac{\partial}{\partial\theta}\left(\frac{1}{\sin\theta}\frac{\partial}{\partial\theta}\right).$$

In the outer flow region $r > a$ we make the analogue of the Stokes ansatz

$$p = -\mu B \frac{\cos\theta}{r^2}\,, \tag{8}$$

$$v_r = \cos\theta \left(q - \frac{B}{r} + \frac{2C}{r^3}\right), \qquad v_\theta = -\sin\theta \left(q - \frac{B}{2r} - \frac{C}{r^3}\right) \tag{9}$$

satisfying the condition (5) at infinity, where

$$q - \frac{B}{2a} - \frac{C}{a^3} = 0 \tag{10}$$

in view of the second condition (4).

In the inner flow region $r < a$ the ansatz

$$\Psi = \sin^2\theta \cdot g(r)$$

for the stream function Ψ in (7) leads to the representation

$$\Psi = \sin^2\theta (br^2 + cr^4)$$

which by (6) yields

$$v_r = 2\cos\theta (b + cr^2)\,, \qquad v_\theta = -2\sin\theta (b + 2cr^2) \tag{11}$$

and by (1)

$$p = 20\mu c r \cos\theta\,. \tag{12}$$

Here in correspondence to (3) and (8) a free additive constant in p is put equal to zero. Further, in view of the first condition (4) there must be

$$b + 2a^2 c = 0\,. \tag{13}$$

The remaining conditions (2), (3) are fulfilled if

$$2(b + a^2 c) = q - \frac{B}{a} + \frac{2C}{a^3} = k\left(20ac + \frac{B}{a^2}\right). \tag{14}$$

Determining the constants B, C, b, c from equations (10), (13), (14), by (8), (9) and (11), (12) we arrive at the following solution:

In $r > a$

$$\left.\begin{aligned} p &= -\frac{3}{2}\mu q a \frac{\cos\theta}{r^2} \cdot \alpha_0\,, \\ v_r &= q\cos\theta\left(1 - \frac{3}{2}\frac{a}{r}\alpha_0 + \frac{1}{2}\frac{a^3}{r^3}\beta_0\right), \\ v_\theta &= -q\sin\theta\left(1 - \frac{3}{4}\frac{a}{r}\alpha_0 - \frac{1}{4}\frac{a^3}{r^3}\beta_0\right) \end{aligned}\right\} \tag{15}$$

with

$$\alpha_0 = \frac{a + 10k}{a + 10.5k}, \qquad \beta_0 = \frac{a + 12k}{a + 10.5k};$$

in $r < a$

$$\left.\begin{aligned} p &= -15\mu q \frac{r}{a^2} \cos\theta \cdot \gamma_0, \\ v_r &= 3q \cos\theta \left(1 - \frac{r^2}{2a^2}\right) \gamma_0, \\ v_\theta &= -3q \sin\theta \left(1 - \frac{r^2}{a^2}\right) \gamma_0 \end{aligned}\right\} \tag{16}$$

with

$$\gamma_0 = \frac{k}{a + 10.5k}.$$

From (15) and (16) we obtain the hydrodynamic pressure difference

$$p^- - p^+ = -\frac{3}{2} \mu \frac{q}{a} \cos\theta \cdot [\alpha_0 - 10\gamma_0]$$

and for the friction pressure p_{rr} and $p_{r\theta}$ the continuity $p_{rr}^- = p_{rr}^+$ and the formula

$$p_{r\theta}^- - p_{r\theta}^+ = \frac{3}{4} \mu \frac{q}{a} \sin\theta \cdot [\alpha_0 + \beta_0 + 8\gamma_0]$$

on $r = a$. This yields the following formula for the drag on the sphere (cf. [1])

$$D_0 = -6\pi\mu a q \cdot \alpha_0, \tag{17}$$

which differs by a factor $\alpha_0 < 1$ given by

$$\alpha_0 = 1 - \frac{0.5k}{a + 10.5k}$$

from the Stokes formula. Further the flow rate through the sphere is

$$Q_0 = \frac{3}{2} \pi a^2 q \cdot \gamma_0 \tag{18}$$

with the above value of γ_0.

The sphere with inner ball

We further deal with the analogous problem for a sphere of radius a with permeable surface containing in its interior a concentric impermeable sphere of radius $b < a$. Then we again have the differential system (1) in the flow regions $r > a$ and $b < r < a$ with the condition (5) at infinity, the boundary conditions (2)–(4) on $r = a$ and additionally the usual condition of adherence

$$v_r(b, \theta) = v_\theta(b, \theta) = 0 \tag{19}$$

on $r = b$.

The boundary value problem (1)–(5), (19) can be solved in quite the same way as above. In $r > a$ the ansatz (8), (9) again holds, whereas in $b < r < a$ the general ansatz for the stream function Ψ

$$\Psi = \sin^2\theta\, (cr^{-1} + dr + er^2 + fr^4)$$

has to be used.

In this way by a somewhat lengthy, but straightforward calculation we get the following solution:

In $r > a$

$$\left.\begin{aligned} p &= -\frac{3}{2} \mu q a \frac{\cos\theta}{r^2} \cdot \alpha, \\ v_r &= q \cos\theta \left(1 - \frac{3}{2}\frac{a}{r}\alpha + \frac{1}{2}\frac{a^3}{r^3}\beta\right), \\ v_\theta &= -q \sin\theta \left(1 - \frac{3}{4}\frac{a}{r}\alpha - \frac{1}{4}\frac{a^3}{r^3}\beta\right), \end{aligned}\right\} \tag{20}$$

where

$$\alpha = 1 - \frac{k}{a}\frac{N_0}{N}, \qquad \beta = 1 + \frac{3k}{a}\frac{N_0}{N}$$

with $N = 2N_0 + 3kN_1$,

$$N_0 = 4a^6 - 9a^5b + 10a^3b^3 - 9ab^5 + 4b^6 = (a - b)^4 [4a^2 + 7ab + 4b^2] > 0,$$

$$N_1 = 28a^5 - 15a^4b - 10a^2b^3 - 3b^5 = (a - b) [28a^4 + 13a^3b + 13a^2b^2 + 3ab^3 + 3b^4] > 0;$$

in $b < r < a$

$$\left.\begin{aligned} p &= -30\mu q \frac{r}{a^2} \cos\theta \cdot \gamma - 6\mu q \frac{b}{r^2} \cos\theta \cdot \delta, \\ v_r &= 3q \cos\theta \left(\varkappa - \frac{r^2}{a^2}\gamma - \frac{2b}{r}\delta + \frac{2b^3}{r^3}\lambda\right), \\ v_\theta &= -3q \sin\theta \left(\varkappa - \frac{2r^2}{a^2}\gamma - \frac{b}{r}\delta - \frac{b^3}{r^3}\lambda\right), \end{aligned}\right\} \tag{21}$$

where

$$\gamma = \frac{ka^2N_2}{N}, \qquad \delta = \frac{kN_3}{N}, \qquad \varkappa = \frac{kN_4}{N}, \qquad \lambda = \frac{ka^2N_5}{N}$$

with

$$N_2 = 4a^3 - 3a^2b - b^3 = (a - b) [4a^2 + ab + b^2],$$

$$N_3 = 6a^5 - 5a^3b^2 - b^5 = (a - b) [6a^4 + 6a^3b + a^2b^2 + ab^3 + b^4],$$

$$N_4 = 8a^5 - 5a^2b^3 - 3b^5 = (a - b) [8a^4 + 8a^3b + 8a^2b^2 + 3ab^3 + 3b^4],$$

$$N_5 = 2a^3 - 3ab^2 + b^3 = (a - b) [2a^2 + 2ab - b^2].$$

In the limit $b \to 0$ there holds $\alpha \to \alpha_0$, $\beta \to \beta_0$, $\gamma \to \gamma_0/2$, $\varkappa \to \gamma_0$.

The drag on the outer sphere $r = a$ is given by

$$D_1 = -6\pi\mu a q \cdot \nu_1 \tag{22}$$

with $\nu_1 = 1 - kM_1/N$,

$$M_1 = 4a^5 + 15a^4b - 10a^2b^3 - 9b^5 = (a - b) [4a^4 + 19a^3b + 19a^2b^2 + 9ab^3 + 9b^4] > 0$$

and the flow rate through the sphere $r = a$ by

$$Q = 3\pi a q k N_0/N. \tag{23}$$

In the limit $b \to O$, of course, D_1 and Q tend to D_0 and Q_0, respectively, whereas in the limit $b \to a$ the value of D_1 tends to $-4\pi\mu a q$, the part of the drag in the Stokes formula coming from the friction pressure (and Q tends to zero).

Finally, the drag on the inner sphere $r = b$ has the value

$$D_2 = -24\pi\mu b q \cdot \nu_2 \tag{24}$$

with $\nu_2 = kM_2/N$,

$$M_2 = 6a^5 - 5a^3b^2 - b^5 = (a - b) [6a^4 + 6a^3b + a^2b^2 + ab^3 + b^4] > 0.$$

In the limit $b \to a$ the value of D_2 tends to $-2\pi\mu a q$, the part of the drag in the Stokes formula coming from the hydrodynamical pressure.

Therefore, if we span a permeable (porous or perforated) hull over an impermeable sphere approximately 2/3 of the Stokes drag acts on the hull and 1/3 of it on the sphere.

References

1 SOMMERFELD, A.: Vorlesungen über Theoretische Physik, Band II: Mechanik der deformierbaren Medien. Akadem. Verlagsgesellschaft Geest u. Portig, Leipzig 1954.
2 JOSEPH, D. D.; TAO, L. N.: The effect of permeability on the slow motion of a porous sphere in a viscous liquid. ZAMM 44 (1964), 361–364.

Received February 29, 1988

Address: Prof. Dr. L. v. WOLFERSDORF, Bergakademie Freiberg, Sektion Mathematik, Akademiestraße 6, DDR-9200 Freiberg

Zeitschrift für Angewandte Mathematik und Mechanik

Volume 69 **1989** **Number 2**

Contents

Zeitschrift für Angewandte Mathematik und Mechanik is indexed in Current Contents/Physical, Chemical & Earth Sciences; Science Citation Index; Mechanics; Applied Mechanics Reviews; Mathematical Reviews; Zbl. Math.

Bei Fragen zur Produktsicherheit wenden Sie sich bitte an:
If you have any questions regarding product safety,
please contact:

Walter de Gruyter GmbH
Genthiner Straße 13
10785 Berlin
productsafety@degruyterbrill.com